SpringerBriefs in Earth Sciences

H0234538

For further volumes:
http://www.springer.com/series/8897

Basudeb Bhatta

Research Methods in Remote Sensing

 Springer

Basudeb Bhatta
Computer Aided Design Centre
Department of Computer Science
and Engineering
Jadavpur University
Kolkata
India

ISSN 2191-5369   ISSN 2191-5377   (electronic)
ISBN 978-94-007-6593-1   ISBN 978-94-007-6594-8   (eBook)
DOI 10.1007/978-94-007-6594-8
Springer Dordrecht Heidelberg New York London

Library of Congress Control Number: 2013934101

Printed on acid-free paper

Springer is part of Springer Science+Business Media (www.springer.com)

Dedicated to research and researchers

Preface

In the early days of remote sensing, concerns of research were primarily ranged over contemporary physical and biological (biophysical) space and their arrangements as they could be documented. The methods that were used to explain, model, and predict different biophysical aspects became progressively more quantitative. Further, the new technologies and theoretical perspectives that emerged in the past few decades helped to redefine the objects of inquiry and extend the methods in use for collecting and analyzing remote sensing data and evaluating researches.

Being a blend of science, art, and technology, and being multidisciplinary in nature, remote sensing generally associates complex nonlinear research methods. Remote sensing has many different sensors and a wide variety of application areas. As a result, the research methods in this emerging field became more complicated and diverse. With the advent of new generation sensors and computer-based techniques for image analysis, remote sensing imageries are now being used more and more in several new folds of scientific researches. Because of its vastness, often, remote sensing becomes a distinct field of study rather than being utilized as a tool in a scientific field. As a result, new researchers in this field often get confused and overlook several issues important to be considered.

This book is an introduction to research methods in remote sensing. A research method is a way of collecting and analyzing the data. This sounds very 'nuts and bolts', but there is no way to properly engage in research (or in methods) without also tackling some of the fundamental theoretical questions. These questions are philosophical in nature, e.g., ontology, epistemology, paradigm, ethics, etc. This book is to furnish the overall concepts of research methods in Remote Sensing; starting from the theoretical ontology to the documentation of research. This book, therefore, covers the theory while providing a solid basis for engaging in concrete research activities. This book is not intended to become a textbook of remote sensing; rather, it has the intention to guide a researcher in conducting their research by documenting the issues that are generally not covered by a textbook.

The book is comprised with eight chapters. Chapter 1 is mainly aimed to document the definitions and overview. It begins with the definition and application areas of remote sensing of the earth's surface, and proceeds toward the research types and research framework in the light of remote sensing. Chapter 2 is intended to discuss the entire research framework—ontology, epistemology, paradigm, methodology, methods, conclusions, and recommendations. Chapter 3 is aimed to discuss the data and their collection/selection methods and related issues. First it discusses the factors influencing the selection of remote sensing data for different types of applications; and then it addresses the ground truth and other ancillary data. Chapter 4 emphasizes the general discussion of remote sensing data analysis. This chapter is based on concepts rather than tools and techniques; constraints and freedoms are also addressed in context. Chapter 5 deals with the research design and its parts—sampling design, observational design, analytical design, and operational design. Chapter 6 helps to understand the nature of power and politics and the critical role of ethics in scientific research, especially remote sensing research. Chapter 7 is aimed to discuss the methods and issues involved in documenting a research outcome. It is a guide on how to write a research paper, dissertation, and thesis.

This book will be of value for the remote sensing researchers from many disciplines. Masters and Ph.D. students of remote sensing will appreciate this book to conduct their researches. This book may help the academicians for preparing lecture notes and delivering lectures. Industry professionals may also be benefited from the discussed methods along with numerous citations. The physical baggage of this book has been kept to a minimum in order to maximize accessibility and readability by a large segment of researchers in the field of remote sensing.

Acknowledgments

I am grateful to all the authors of the numerous books and research publications mentioned in the list of references at the end of each chapter. These valuable literatures formed the foundation of this book. I express my gratitude to those teachers, researchers, and organizations for their contributions that reinforced my knowledge.

I would then like to express my profound gratitude to Prof. Rana Dattagupta, Former Director, CAD Centre, Jadavpur University, and Prof. Sivaji Bandyopadhyay, Director, CAD Centre, Jadavpur University, for extending all possible facilities to write this book.

I am very much thankful to my colleagues, especially Mr. Biswajit Giri, Mr. Chiranjib Karmakar, Mr. Subrata Das, Mr. Santanu Glosal, and Mr. Uday Kumar De. Without their help and cooperation writing this book was never possible.

I would like to express my gratitude to my parents who have been a perennial source of inspiration and hope for me. I also want to thank my wife Chandrani, for her understanding and full support, while I worked on this book. My little daughter, Bagmi, deserves a pat for bearing with me during this rigorous exercise.

Basudeb Bhatta

Contents

Chapter 1
Introduction to Remote Sensing and Research

Abstract This chapter is mainly aimed to document the definitions and overview, as it is expected from any standard book. The chapter begins with the definition and application areas of remote sensing of the earth, and proceeds towards the research types and research framework in the light of remote sensing. Research framework will be elaborated later in Chap. 2; however, this chapter will act as a prelude. Remote sensing, being a blend of science, technology, and art, and being multidisciplinary in nature, is different from other fields of study. Therefore, the research methods in remote sensing, nested within the theoretical coordinates of research framework, are necessary to be considered and understood separately.

Keywords Remote sensing • Research • Philosophy • Science • Technology • Art • Hypothesis • Theory • Law • Scientific thinking • Empiricism • Rationalism • Logical reasoning • Scepticism • Research types • Research framework

1.1 Remote Sensing

Remote sensing has as many definitions as its applications. Perhaps, the simplest definition of remote sensing is "acquiring of data about an object without touching it". Although it is short, simple, and memorable, this definition is extremely vague. Unfortunately, it excludes little from the province of remote sensing of the earth's surface. It encompasses virtually all remote sensing devices, including cameras, telescopes, optical-mechanical scanners, linear and area arrays, lasers, radio-frequency receivers, radar systems, sonar, seismographs, gravimeters, magnetometers, X-ray, and other medical applications (Bhatta 2011).

For our purposes, in this text, we are going to restrict our discussion to earth observation from overhead sensors mounted on aircrafts and satellites. Given this restriction, a narrow definition of remote sensing can be given as "the non-contact recording of information from the ultraviolet, visible, infrared, and microwave regions of the electromagnetic spectrum by means of instruments such as cameras, scanners, lasers, linear arrays, and/or area arrays located on platforms such as aircraft or spacecraft, and the analysis of acquired information by means of visual and digital image processing" (Jensen 2006).

B. Bhatta, *Research Methods in Remote Sensing*, SpringerBriefs in Earth Sciences, DOI: 10.1007/978-94-007-6594-8_1, © The Author(s) 2013

Remote sensing, so far as our interests are concerned, is relatively a new field. It is important to realize, many of the principles in this young field are still being formulated, and in many areas the consistent structure and terminology we expect in more mature fields may be lacking. On the positive side, remote sensing as a relatively young field, offers a myriad of opportunities for exploring unanswered (and often as yet unasked) questions. These questions address what is to be learned about the earth's surface and about the earth's land, water, and atmosphere. They can be further extended to include the living agents in the earth.

1.1.1 Research and Application Areas of Remote Sensing

Remote sensing is a technique that can be used in a wide variety of disciplines, but is not a discipline or subject by itself. Since remote sensing is developing itself at a rapid rate, several applications are being tested and several more remain to be tested. Several applications have already been tested in which some proved effective, while others have not succeeded. However, there can be almost endless applications of remote sensing techniques to tackle problems related to land surface, sea surface, and atmospheric features and processes. Table 1.1 lists most common research areas of remote sensing. However, there are still many research areas that have not been documented in this table.

1.1.2 Whether Remote Sensing is Science, Art, or Technology

'Why should we consider remote sensing data' or 'what are the advantages of it' or 'what are the properties associated with it' can be found in any textbook. In the context of this book, we should rather raise the question whether remote sensing is science, or technology, or art. Because, the methodology involved in research may vary extensively among these three. Many of the literature preferred to define remote sensing as "science and art of obtaining and interpreting information about an object, area, or phenomenon through the analysis" (e.g., Jensen 2006; Bhatta 2011). However, remote sensing is a perfect blend of science, technology, and art. Lillesand et al. (2007) stated "Remote sensing is the science, technology, and art of obtaining information about an object, area, or phenomenon by analyzing data acquired by a device that is not in physical direct contact with the object, area or phenomenon under investigation". Alavipanah et al. (2010) have shown a conceptual diagram (Fig. 1.1) of blending science, technology, and art as remote sensing.

Science is a system of acquiring knowledge based on the scientific methods, as well as the organized body of knowledge gained through such research. It is the understanding and continuous exploration of the natural world. Science is often driven by whim or curiosity without having any application goal. Science, as defined here, is sometimes termed pure science to differentiate it from applied

Table 1.1 Common research areas of remote sensing (Bhatta 2011)

Land-use/land-cover	• Natural resource management
	• Wildlife habitat protection
	• Baseline mapping for geographic information system (GIS) input
	• Urban expansion/encroachment
	• Routing and logistics planning for seismic/exploration/resource extraction activities
	• Damage delineation (tornadoes, flooding, volcanic, seismic, fire, and terrorist activities)
	• Legal boundaries for tax and property evaluation
	• Target detection—identification of landing strips, roads, clearings, bridges, and land/water interface
Agriculture	• Classification of crop type
	• Assessment of crop condition
	• Estimation of crop yield
	• Mapping of soil characteristics
	• Mapping of soil management practices
	• Compliance monitoring (farming practices)
Forestry	• Reconnaissance mapping
	Forest-cover type discrimination
	Agroforestry mapping
	• Commercial forestry
	Clear-cut mapping/regeneration assessment
	Burn delineation
	Infrastructure mapping/operations support
	Forest inventory
	Biomass estimation
	Species inventory
	• Environmental monitoring
	Deforestation (rainforest, mangrove colonies)
	Species' inventory
	Watershed protection (riparian strips)
	Coastal protection (mangrove forests)
	Forest health and vigour
Geology	• Mapping of surficial deposit/bedrock
	• Lithological mapping
	• Structural mapping
	• Lineament extraction
	• Exploration/exploitation of sand and gravel (aggregate)
	• Mineral exploration
	• Exploration of hydrocarbon
	• Environmental geology
	• Geobotany
	• Baseline infrastructure
	• Mapping and monitoring
	• Event mapping and monitoring
	• Geo-hazard mapping
	• Planetary mapping

(continued)

Table 1.1 (continued)

Geomorphology	• Endogenetic processes 　Volcanism 　Plate tectonics 　Diastrophism: folding, faulting, and warping • Exogenetic systems 　Weathering 　Mass wasting 　Erosion, transportation, and depositional processes 　Alluvial/fluvial (flowing water) 　Glacial (ice) 　Eolian (wind) 　Coastal (waves)
Urban	• Pattern/process • Growth/sprawl • Mapping • Planning, simulation and modelling
Hydrology	• Wetlands mapping and monitoring • Water quality monitoring • Soil moisture estimation • Snow pack monitoring/delineation of extent • Measuring snow thickness • Determining snow–water equivalence • River and lake ice monitoring • Flood mapping and monitoring • Glacier dynamics monitoring (surges and ablation) • River/delta change detection • Drainage basin mapping and watershed modelling • Irrigation canal leakage detection • Irrigation scheduling
Cartography and mapping	• Planimetry • Digital elevation models • Baseline thematic mapping/topographic mapping
Ocean and coastal	• Ocean pattern identification 　Currents, regional circulation patterns, and shears 　Frontal zones, internal waves, gravity waves, eddies, 　up-welling zones, and shallow water bathymetry • Storm forecasting 　Wind and wave retrieval 　Fish stock and marine mammal assessment 　Water temperature monitoring 　Water quality 　Ocean productivity, phytoplankton concentration, and drift • Oil spill 　Mapping and predicting oil spill extent and drift 　Strategic support for oil spill emergency response decisions 　Identification of natural oil seepage areas for exploration • Shipping 　Navigation routing 　Traffic density studies

(continued)

Table 1.1 (continued)

Operational fisheries surveillance
Near-shore bathymetry mapping
• Inter-tidal zone
Tidal and storm effects
Delineation of the land/water interface
Mapping shoreline features/beach dynamics
Coastal vegetation mapping
Human activity/impact
• Sea ice
Ice concentration
Ice type/age/motion
Iceberg detection and tracking
Surface topography
Tactical identification of leads, navigation, and safe shipping routes/rescue
Ice condition (state of decay)
Historical ice and iceberg conditions and dynamics for planning purposes
Wildlife habitat
Pollution monitoring
Meteorological/global change research
Atmospheric applications

Fig. 1.1 Three main dimensions of remote sensing (Alavipanah et al. 2010)

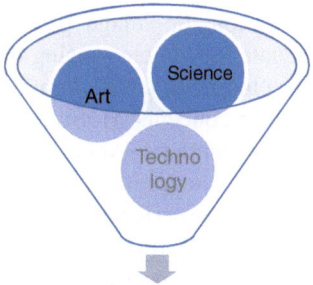

Remote Sensing

science that is the application of scientific research to specific human needs. Science refers to a system of acquiring knowledge. This system uses observation and experimentation to describe and explain natural phenomena. Technology is applying the outcome of scientific principles to innovate and improve the man-made things in the world. The output of Technology is a new or better process of doing. In human society, it is a consequence of science. Art is the product or process of deliberately arranging items in a way that influences and affects one or more of the senses, emotions, and intellect. In simple words, art is the expression or application of human creative skill and imagination. If one can use different process to create a thing (output) using the same inputs it is called art. Generally, in science or technology, we use a standard process to create a thing (output) using same inputs. Science becomes art when one crosses the boundaries of set rules or explicit instructions and run on instinct or intuition. It is much more evident in areas of science that have not yet been fully discovered.

Remote sensing is a tool or technique similar to mathematics. Using sophisticated sensors to measure the amount of electromagnetic energy exiting an object or geographic area from a distance, and then extracting valuable information from the data using mathematically and statistically based algorithms is a scientific and technologic activity. It functions in harmony with other spatial data collection techniques or tools of the mapping sciences, including cartography and GIS. The synergism of combining scientific knowledge with real-world analyst experience allows the interpreter to develop heuristic rules of thumb to extract valuable information from the imagery. It is a fact that some image analysts are much superior to others because they: (1) understand the scientific principles better, (2) are more widely travelled and have seen many landscape objects and geographic areas first hand, and (3) can synthesize scientific principles and real-world knowledge to reach logical and correct conclusions (Jensen 2006).

Interpreting remotely sensed images is an open-ended task (Hoffman and Markman 2001). The perception of image in the part of the interpretation of remotely sensed images are the most outstanding and artistic parts of remote sensing (Fig. 1.2). Human is created such that he is able to percept the realities of the entity. In other words, human is equipped with intellect by which he can percept his surrounding world.

Automatic image processing techniques (by using computers) remain inadequate for remote sensing data analysis (Fried et al. 1988). The human must be in the 'loop'; since the human, unlike the computer, can perceive and can form and reform concepts (Hoffman and Markman 2001). Important to realize, human interpreter can derive very little information using a point-by-point approach. Many of original interpretations depended not only on the imagery itself but also on the skill and experience of interpreter (Campbell 1996).

For the purpose of the perception of image in interpretation of remotely sensed images, the necessity of using artistic outcomes and in particular applied arts

Fig. 1.2 Elements of image understanding (Alavipanah et al. 2010)

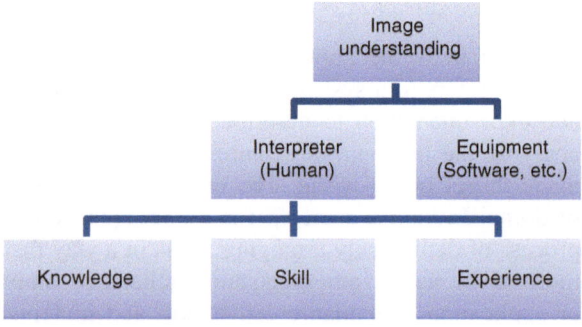

become more prominent. Having ability to make a visual inspection along with 'visual knowledge', beautiful selection and the efficiency of colours by considering the principles of compatibility and lack of compatibility of colour, increase of idea fertilization and ability to have a specific observance with the help of imagination and mental creativity and order are among the consequences which make possible utilization of this issue and having access to that will increase the ability to interpret (Alavipanah et al. 2010). Since the visual interpretation of remotely sensed images is mostly accompanied with individual judgment, a researcher should know how to employ the scientific and proper methods to reach the goal. In most of the cases, the conditions of the earth which appear in the image are complex. As a result, sometimes, knowledge and experience of an interpreter fails to make a link between the phenomena of the earth and the information content of an image.

Therefore, from the preceding discussion, it is evident that remote sensing is a blend of science, technology, and art. The important thing one should realize is that information extracted from remote sensing data may vary from analyst to analyst to some extents and achieving 100 % accuracy is never possible.

1.2 Which Method: Scientific, Technological, or Artistic

Which research method to be followed in remote sensing? Is the research method involved in remote sensing is a blend of scientific, technological and artistic methods? Before answering these questions let us first have a brief discussion on scientific, technological and artistic methods.

1.2.1 Scientific Method

The scientific method is a body of techniques for investigating phenomena, acquiring new knowledge, or correcting and integrating previous knowledge (Goldhaber and Nieto 2010). To be termed scientific, a method of inquiry must be based on empirical and measurable evidence subject to specific principles of reasoning (Newton 1999). The Oxford English Dictionary says that the scientific method is "a method or procedure that has characterized natural science since the 17th century, consisting in systematic observation, measurement, and experiment, and the formulation, testing, and modification of hypotheses". In simple words, scientific method is a way to ask and answer scientific questions by making observations and doing experiments. The scientific method is a way to make sure that our experiment can give a good answer to our specific question. It is a logical and rational order of steps by which scientists come to conclusions about the world around them.

Scientific method is not much different from our day-to-day ways of learning about the world (Kosso 2011). Without really thinking about the steps or the standards, common sense invokes the same process of evidence and reasoning as scientists more explicitly follow. The key difference between science and our daily life is that the scientific process is more deliberate and explicit in following the steps and standards of the method. We do it tacitly in our daily life, but in science the procedures must be articulated and described. "The scientific process is purposefully slowed down in the interest of control and transparency. A single step of gathering evidence may take days, or weeks, or even years. There are no snap decisions like the ones we are forced to make in crossing the street or preparing breakfast. The components and conduct of the scientific process are essentially similar to those of common sense, but things proceed more slowly and deliberately in science" (Kosso 2011).

Science is also distinguished by a greater dedication to the results of the method than is characteristic of our daily life. Common sense is, alas, not as common as it should be. Among friends we seem to accommodate such weaknesses as wishful thinking, superstition, and stereotyping. Among scientists we do our best to avoid these violations of the method, following the evidence and the logic as best we can, regardless of our preconceptions. Science is thus more deliberate and dedicated than non-science in following the method. It is also more public and open to independent review. In life we are rarely asked to actually produce the evidence and reasons for our knowledge claims. We trust each other and ourselves, unless the claim is surprising. But in science, one is routinely expected to not just have good reasons in support of one's claims but to actually produce those reasons.

Science is a method of investigating nature—a way of knowing about nature—that discovers reliable knowledge about it. In other words, science is a method of discovering reliable knowledge about nature. Considering this statement, we must admit, knowledge is the goal of science. Basic research seeks reliable knowledge, and applied research seeks useful knowledge. But if knowledge were our primary goal as scientists, we would spend much of our available time in reading the literature rather than in slowly gathering new data. "Science is not static knowledge; it is a dynamic process of exploring the world and seeking to obtain a trustworthy understanding of it. Everyone practices this process, to some extent. Science is not the opposite of intuition, but a way of employing reality testing to harness intuition effectively and productively" (Jarrard 2001).

All scientific knowledge must be based on observation. The word 'observation' has dual meaning in remote sensing—(1) observation of the earth-surface by using remote sensing sensors/cameras, and (2) observation of earth-surface features (their shapes, sizes, spatial relationships, spectral reflectance characteristics, and so on) on the remote sensing images by the researcher. Other than these two, ground-truthing or ground observation is also made by the researcher. However, ground activities, in remote sensing, are generally considered as 'truth' rather than simple observations. They are made to establish a link between the remote sensing imagery and the real world. Coming back to the scientific observation, it is the basis of scientific method, but there is some ambiguity in how close a link is required between observation and theory. The method cannot be simply a process of generalizing knowledge from observations, since some, at least tentative, knowledge is prerequisite for making scientific observations (Kosso 2011). Clarity on the scientific method begins with clarity on the key terms and concepts. Three such terms in particular do a lot of work in most descriptions and evaluations of science: (1) *hypothesis*, (2) *theory*, and (3) *law*.

Hypothesis is a logical but unproven explanation for a given set of facts. It is used as a starting point for further experimentation and observation. A hypothesis must be testable; otherwise it is a worthless hypothesis. A hypothesis is tested by comparing results of experiments with the hypothesis' predictions (Anderson 2000). Unlike 'theory' (explained later) the term 'hypothesis' does refer to the amount of good reason to believe, that is, to the location on the spectrum between certainty and speculation. A hypothesis is a thing that has little testing and is consequently located near the speculation-end of the spectrum (Kosso 2011). It is a theory for which the connection to fact is unknown or unclear, but usually there is some tentative reason to believe this link will be made. There is reason to think that evidence and support from other theories will allow the hypothesis to move up the spectrum to the well-supported side.

Theory is a hypothesis which has been tested numerous times and found to explain previous observations and make accurate predictions about future observations. The division between hypothesis and theory is a bit fuzzy, but theory is sort of when there is no more 'reasonable doubt' of the hypothesis' truth. However, Kosso (2011) stated that sometimes theory is used in a pejorative way, as an invitation to doubt or even believe the opposite, as in 'that's just a theory'. But other times theory is an honour, implying a coherent network of ideas that successfully explain some otherwise mysterious aspect of nature. A clear, unambiguous meaning of the term theory emerges from a survey of examples of scientific theories and seeing what it is they all have in common, what it is that makes them theoretical. It is not going to have anything to do with how well-tested or well-confirmed an idea is, or how likely it is to be true. A theory is true if it describes unobservable things that really exist, and describes them accurately. Otherwise it is false. This shows the mistake in contrasting theory and fact. A fact is an actual state of affairs in nature, and a theory, or any statement for that matter, is true if it matches a fact (Kosso 2011). Some theories are true, some are false, and the scientific method is what directs us in deciding which are which.

Laws are theories which have been extensively tested and have never been disproven in any test. An example is 'fire produces heat'. Theories differ with laws in terms of their generality. The *big bang theory*, for example, is about a singular, unique event. It is not general at all, despite being about the entire universe. The *theory of gravity* is very general. It is about all objects with mass and their resulting attraction. The most general theories are laws, for example, the theory of gravity is a law. In other words, laws are theories of a particular kind, the ones that identify whole categories of things and describe their relations in the most general terms (Kosso 2011).

Figure 1.3 explains the process of scientific method, which shows how an observation forms a law. Just think about Sir Isaac Newton's observation about the apple and his law of gravity.

Fig. 1.3 Scientific method
(Anderson 2000)

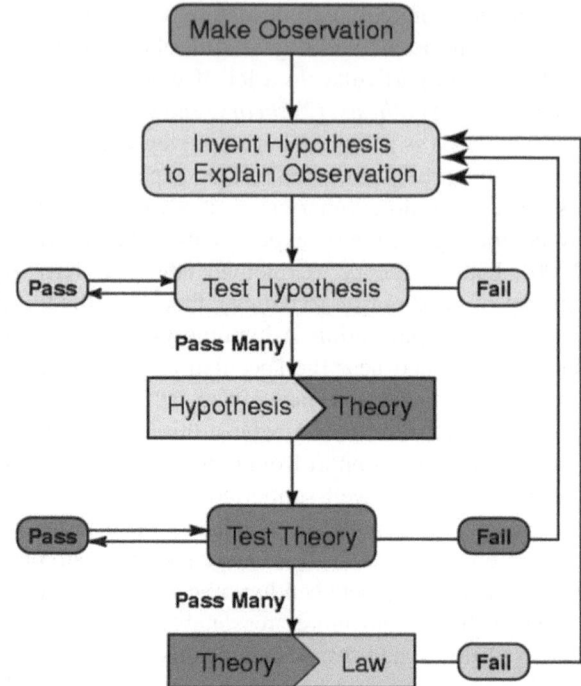

1.2.2 Technological Method

Technological research is an applied research, oriented toward engineering disciplines (but not to a specific product or process) and aimed at developing tools and test equipment and procedures, and at providing solutions to specific technical problems. The words science and technology can and often are used interchangeably. But the goal of science is the pursuit of knowledge for its own sake while the goal of technology is to solve problems and improve human life. In that sense, science is about *knowing* and technology is about *doing*. Technology can be thought as the practical application of science. Often it is said 'technology starts where science concludes'. Technology is always dependent on science; because it is the application of knowledge attained through science to modify human life. For example, quantum theory (science) enabled our electronic devices (technology). Science is also often dependent on technology; for example, scientific instruments (technology) are often needed to make measurements to confirm theories (science). Science and technology are often overlapping and mutually dependent on each other to some extent. Applied science and technology are actually hard to distinguish.

Cellular telephones, computers, medical lasers, disease-resistant crops, satellites, biotechnology, optical fibre networks—all these twentieth-century technologies and many others can trace their origins at least in part to both science and technological research. New knowledge alone is not enough to achieve major economic, military, or social objectives. But through the combined efforts of business, government, and academic and other non-profit organizations, new knowledge has been converted into new technologies, new means of production, and new industries. In this process, technological research has enhanced national security, improved human health, produced a stronger economy, and led to a cleaner environment. Technological research will be even more influential in the twenty-first century than it has been in the twentieth century. No one can predict which technologies will define the next century. But we know that the increasing interconnection of computers into a global network will transform work, communications, entertainment, and education.

Technological research method does not start with observation, and then forming the hypothesis and testing hypothesis as with the scientific method. Rather it starts with defining the research problem and research goal, and then experimenting and testing to achieve the defined goal. The outcome of a technological method is not a theory or law rather new/improved tools and test equipments and models/procedures. Therefore, technological method is a straightforward process.

1.2.3 Artistic Method

One of most intriguing aspects about art today is its entanglement with theory. In fact, contemporary art practice is now so highly saturated with theoretical knowledge that it is becoming a research practice in and of itself. Artists have not only taken up art criticism and negotiations, they now also integrate research methods and scientific knowledge into their artistic process to such a degree that it even seems to be developing into an independent form of knowledge on its own (Busch 2009). For example, photographers should have knowledge of optics related to their domain.

Artistic appropriation of knowledge evokes different, independent forms of knowledge, in order to complement scientific research with artistic research. With regard to the relationship between philosophy and art, this implies that artistic practice is more than just an application of theory and that theory is more than a mere reflection on practice. Deleuze (1977) perceived this unique relationship as "a system of relays within a larger sphere, within a multiplicity of parts that are both theoretical and practical". Art and theory, in effect, are nothing more than two different forms of practice interrelated through a system of interaction and transferences. In this constellation, philosophy neither brings the arts to the point nor does art sensualise philosophical truths; philosophy serves a knowledge-based artistic practice as a point of reference, similar, conversely, to how art might affect theoretical practice.

An art method differs from a science method, perhaps mainly insofar as the artist is not always after the same goal as the scientist. In art it is not necessarily all about establishing the exact truth so much as making the most effective form (painting, drawing, poem, novel, performance, sculpture, etc.) through which ideas, feelings, perceptions can be communicated to a public. With this purpose in mind, some artists will exhibit preliminary sketches and notes which were part of the process leading to the creation of a work. Sometimes, in conceptual art, the preliminary process is the only part of the work which is exhibited, with no visible end result displayed. In such a case the 'journey' is being presented as more important than the destination.

The most important thing we need to understand that artistic method does not establish a theory or law, or develop a tool or test equipment or model/procedure. It produces a 'creation', most often unique. The creation, whether appreciated or criticized, remains as a piece of work. It cannot be rejected and dumped since it is not dependent on establishing the 'truth'.

1.2.4 Remote Sensing Research Method

Coming back to the question of our interest: which method to be followed in remote sensing? The discussions in the preceding sections suggest that the remote sensing method is either scientific or technologic or a combination of both. Scientific method is based on observations—observation of spectral reflectance characteristics of vegetation for example—and drawing conclusion that vegetation gives highest reflectance in the near-infrared band and lowest reflectance in the red band of optical electromagnetic spectrum. Now the technology comes to develop tools, models, or procedures, or to bring a solution to human problem; for example, developing a model to simulate future urban growth. However, the basic questions posed in the field of remote sensing are of scientific in nature. Remote sensing, although, follows art method to interpret the imagery or to infer indirect relationships and thereby information (intuitive knowledge), the research is not conducted in artistic method because it requires establishing the 'truth' rather than producing a 'creation'. Therefore, our further discussions will be based on the scientific method and, where required, on technological method.

1.3 Scientific and Critical Thinking

When one uses the scientific method to study or investigate nature or the universe, one is practicing scientific thinking. All scientists practice scientific thinking, of course, since they are actively studying nature and investigating the universe by using the scientific method. But scientific thinking is not reserved solely for scientists. Anyone can think like a scientist who learns the scientific method and, most importantly, applies its precepts, whether he or she is investigating nature or not. When one uses the methods and principles of scientific thinking in everyday life— such as when studying history or literature, investigating societies or governments, seeking solutions to problems of economics or philosophy, or just trying to answer personal questions about oneself or the meaning of existence—one is said to be practicing critical thinking. Critical thinking is a type of reasonable, reflective thinking that is aimed at deciding what to believe or what to do (Ennis 1987). It is a way of deciding whether a claim is always true, sometimes true, partly true, or false. Critical thinking is thinking correctly for oneself that successfully leads to the most reliable answers to questions and solutions to problems. In other words, critical thinking gives you reliable knowledge about all aspects of our life and society, and is not restricted to the formal study of nature. Scientific thinking is identical in theory and practice, but the term would be used to describe the method that gives you reliable knowledge about the natural world. Clearly, scientific and critical thinking are the same thing, but where one (scientific thinking) is always practiced by scientists, the other (critical thinking) is sometimes used by others and sometimes not. Some professionals in the humanities, social sciences, jurisprudence, business, and journalism practice critical thinking as well as any scientist, but many do not. Scientists must practice critical thinking to be successful, but the qualifications for success in other professions do not necessarily require the use of critical thinking, a fact that is the source of much confusion, discord, and unhappiness in our society.

At this point, it is customary to discuss questions, observations, data, hypotheses, testing, and theories, which are the formal parts of the scientific method, but these are not the most important components of the scientific method. The scientific method is practiced within a context of scientific thinking, and scientific (and critical) thinking is based on three primary things: using empirical evidence (empiricism), practicing logical reasonsing (rationalism), and possessing a sceptical attitude (scepticism) about presumed knowledge that leads to self-questioning, holding tentative conclusions, and being undogmatic (willingness to change one's beliefs) (Schafersman 1997). These three ideas or principles are universal throughout science; without them, there would be no scientific or critical thinking. Let us examine each in turn.

1.3.1   Empiricism: The Use of Empirical Evidence

Empirical evidence is evidence that one can see, hear, touch, taste, or smell; it is evidence that is susceptible to our senses. Empirical evidence is important because it is evidence that others besides the researcher can experience, and it is repeatable, so empirical evidence can be checked by the researcher and others after knowledge claims are made by an individual. Empirical evidence is the only type of evidence that possesses these attributes and is therefore the only type used by scientists and critical thinkers to make vital decisions and reach sound conclusions. We can contrast empirical evidence with other types of evidence to understand its value; for example, hearsay evidence is what someone says they heard another say. These are not reliable because one cannot check its source. Empirical evidences, in science, are determined through observations and/or experimentations.

The most common alternative to empirical evidence is *authoritarian evidence*; that means, authorities (people, books, journals, television, newspapers, etc.) tell us to believe. Sometimes, if the authority is reliable, authoritarian evidence is reliable evidence, but many authorities are not reliable (do we believe everything shown in television?), so the researcher must check the reliability of each authority before accepting its evidence. Importantly, one must be their own authority and rely on their own powers of critical thinking to know if what they believe is reliably true. But remember: some authoritarian evidence and knowledge should be validated by empirical evidence, logical reasoning, and critical thinking before one should consider it reliable, and, in most cases, only researcher can do this for her/him-self. It is, of course, impossible to receive an adequate education today without relying almost entirely upon authoritarian evidence. Teachers are generally considered to be reliable and trustworthy authorities, but even they should be questioned on occasion. However, teachers cannot be referred in scientific research; authoritarian evidence must be in recorded form—written or taped or photographed, etc. The most reliable authoritative knowledge in science is, perhaps, peer reviewed journal.

Another name for empirical evidence is *natural evidence*: the evidence found in nature. Naturalism is the philosophy that says that reality and existence (i.e., the universe, cosmos, or nature) can be described and explained solely in terms of natural evidence, natural processes, and natural laws (Schafersman 1997).

1.3.2 Rationalism: The Practice of Logical Reasoning

Scientists and critical thinkers always use logical reasoning. Logic allows us to reason correctly, but it is a complex topic and not easily learned; many books are devoted to explaining how to reason correctly, and we cannot go into the details here. However, it is very important to mention that most individuals do not reason logically, because they have never learned how to do so. Logic is not an ability that humans are born with or one that will gradually develop and improve on its own, but is a skill or discipline that must be learned within a formal educational environment and can be enhanced by continuous practice. Emotional thinking, hopeful thinking, and wishful thinking are much more common than logical thinking, because they are far easier and more congenial to human nature. Most individuals would rather believe something is true because they feel it is true, hope it is true, or wish it was true, rather than deny their emotions and accept that their beliefs are false.

Often the use of logical reasoning requires a struggle with the will, because logic sometimes forces one to deny her/his emotions and face reality, and this is often painful. But remember, emotions are not evidence, feelings are not facts, and subjective beliefs are not substantive beliefs. Every successful scientist and critical thinker spent years learning how to think logically, almost always in a formal educational context. Some people can learn logical thinking by trial and error, but this method wastes time, is inefficient, is sometimes unsuccessful, and is often painful.

1.3.3 Scepticism: Possessing a Sceptical Attitude

The final key idea in science and critical thinking is scepticism, the constant questioning of your beliefs and conclusions. Good scientists and critical thinkers constantly examine the evidence, arguments, and reasons for their beliefs. Self-deception and deception of oneself by others are two of the most common human failings. Self-deception often goes unrecognized because most people deceive themselves. The only way to escape both deceptions by others and the far more common trait of self-deception is to repeatedly and rigorously examine our basis for holding our beliefs. We must question the truth and reliability of both the knowledge claims of others and the knowledge we already possess. One way to do this is to test our beliefs against objective reality by predicting the consequences or logical outcomes of our beliefs and the actions that follow from our beliefs. If the logical consequences of our beliefs match objective reality—as measured by empirical evidence—we can conclude that our beliefs are reliable knowledge (that is, our beliefs have a high probability of being true).

Many people believe that sceptics are closed-minded and, once possessing reliable knowledge, resist changing their minds—but just the opposite is true (Schafersman 1997). A sceptic holds beliefs tentatively, and is open to new evidence and rational arguments about those beliefs. Sceptics are undogmatic, i.e., they are willing to change their minds, but only in the face of new reliable evidence or sound reasons that compel one to do so. Sceptics have open minds, but not so open, they resist believing something in the first place without adequate evidence or reason, and this attribute is worthy of emulation. Science treats new ideas with the same scepticism: extraordinary claims require extraordinary evidence to justify one's credulity. We are faced every day with fantastic, bizarre, and outrageous claims about the natural world; if we don't wish to believe every pseudoscientific allegation or claim of the paranormal, we must have some method of deciding what to believe or not, and that method is the scientific method which uses critical thinking.

1.4 Research and its Types

Research can be defined as the scientific search for knowledge, or as any systematic investigation, to establish novel facts, solve new or existing problems, prove new ideas, or develop new theories, usually using a scientific method. Simply speaking the purpose of research is to find a solution to a particular type of problem or showing the problem a direction towards solution or putting some light in the unknown areas of knowledge for our enlightenment to that particular area of knowledge.

Like every system of this world, research also follows certain rules and takes the help of certain organized procedures in order to get the intended results. In other words, a research works on its methods to serve its true purpose in any area of the knowledge domain. Every research works either to identify new opportunities for us or to give us novel ideas. Research helps us in diagnosing any known problems or opportunities, and to establish a standard of taking action on any chosen area of the knowledge domain. It evaluates and develops the current technologies and systems.

Research may be of different types because the areas of study are different and our problems are also different. Classification of research can also be done from several perspectives. Important to realize, different types of research are often interlinked in scientific work. In the course of the development of scientific study, there seems to have been a constant swing from one of these kinds of research to other. They are mutually supporting and one complements the knowledge necessary for the other kind of research.

Perhaps, the most frequently documented classification of research is *quantitative* and *qualitative*. Quantitative research is based on measurements (in numbers). On the other hand, qualitative research is based on simple observations, not measured numbers. Quantitative research seeks to convert observations into numbers, and explanation/analysis through statistical/mathematical formulas. A type of questions asked, in this type of research, often describes variables, examines relationships among variables, and determines cause-and-effect interactions among variables. One example is "establishing a relation between the vegetation healths and *normalized differential vegetation index* values". Qualitative research emphasizes verbal descriptions and explanations of behaviour. The primary tools for gaining information include observations and interviews. An example is "how to infer the land-use from land-cover information".

Another classification of research is *exploratory*, *explanatory*, and *empirical*. Exploratory (also called *descriptive*) research emphasizes the accurate description of some aspects. A researcher assesses specific characteristics of individuals, groups, objects, situations, or events by summarizing the observations. The descriptive research is directed toward studying 'what' and how many of this 'what'. An example may be "what is the difference in reflectance values between healthy grass and healthy mangrove in the near infrared band during the winter season". This type of research may generate a novel idea in the domain of knowledge. It is primarily done for the purpose of finding anything new in any subject area and always tries to shed some light in the unknown domain of knowledge. This kind of research also helps us to generate new discipline in sciences, and helps us to identify problems of those particular research areas.

The primary goal of explanatory research is to understand or to explain relationships. It uses correlations to study relationships between dimensions or characteristics of individuals, groups, objects, situations, or events. Explanatory research explains how the parts of a phenomenon are related to each other. This type of research asks the question 'why'. One example can be given as "why the healthy grass gives higher reflectance than the healthy mangrove in the near infrared band during the winter season".

Empirical (also called *predictive*) research goes beyond explanation to the prediction of precise relationships between dimensions or characteristics of a phenomenon or differences between groups. In the empirical research one observes or tests real-life data or analyzes the pattern of some specific events in order to identify the nature or the class of trend that specific phenomenon maintains. Based on the test results, researchers try to predict the result of that type of incidents with certain level of confidence. Often this type of research simulates the future based on the observations of past and present; for example, "what may be the built-up pattern in a habitat after 50 years".

Another classification of research is *basic* (also called *pure*) and *applied*. The primary purpose for basic research is discovering, interpreting, and the development of methods and systems for the advancement of human knowledge on a wide variety of scientific matters of our world and the universe. It is often driven by 'whim' or 'curiosity'. It focuses on understanding the phenomena of interest. This type of research is conducted to accumulate information, extending the base of knowledge in a discipline to improve understanding, or to formulate a theory. It is appropriate for discovering general principles of behaviours of earth-surface features and biophysical processes on the earth's surface. One example is "identifying the reflectance characteristics of vegetation in the different regions of electromagnetic spectrum". What for this knowledge will be used is not always defined in such cases. It emphasizes increase in knowledge, not applications of knowledge.

Applied research is a form of systematic inquiry involving the practical application of knowledge gained in basic research. It accesses and uses some part of the research communities' (the academy's) accumulated theories, knowledge, methods, and techniques, for a specific, often state, business, or client driven purpose. Applied research deals with solving practical problems. It focuses on finding an immediate solution to an existing problem. It is designed to indicate how the basic principles can be used to solve problems. This is mainly done by many technological corporate in order to find new/alternative solutions to any particular crisis or problems. An example is "can we propose a better index/formula/method to analyze the vegetation health from remote sensing data". Another simple example is "investigation of forest health in a particular area from remote sensing data". These researches require the knowledge of "reflectance characteristics of vegetation in the different regions of electromagnetic spectrum" that was gained in the basic research.

On the basis of an analysis of different types of research, Ziemski (2005) identified two main kinds: research with a *general intention* and research with an *individual intention*. Research of the first kind aims at establishing the general regularities of sets of objects, their cause and effect relationship, correlation of their characteristics, their independent connection. Research of the second kind aims at identifying individual objects, individual processes; at providing a genetic explanation; at defining their significance for larger units; at predicting their development. Research of first kind may be called *generalizing*, i.e., discovering and accounting for general regularities. Research of the second kind is called *diagnostic* in face of the fact that the explanation of individual objects, their description and explanation is in the nature of a diagnosis.

The United Nations Educational Scientific and Cultural Organization (UNESCO) attempted to classify the research in a different way. Table 1.2 presents their classification of research.

Whatever the type of research we may follow, it should always add new knowledge through: (1) proving new fact that was not known before, (2) validating (or invalidating) the results of previous research, (3) testing of theories, (4) explaining findings of a previous research, and (5) finding new relationships among present phenomena. Research-results should be liable to *testing* (when another researcher chooses the same problem and follows the same steps, she/he probably gets the same results) and *generalization* (i.e., the results could be generalized from the study sample to the whole class). Finally, the research must be ethical; i.e., it does not violate the rights of patients, profession, community, or the researcher her/him-self.

1.4.1 Research Framework

To undertake the stated types of research, there can be a number of ways or approaches that one researcher may follow—these are called research methods. Research method involves collection and analysis of data. Collection and analysis of data is not as simple as it sounds. It involves some fundamental theoretical questions. As Gomez and Jones (2010) discussed, these questions are philosophical and concern the nature of reality (ontology) and how we go about understanding it (epistemology). Such philosophical concerns tend to get stored out into distinct concepts (paradigms) that a researcher can follow as a part of her/his everyday scientific practices. Nested within the theoretical coordinates of paradigms, a set of decisions one has to make about the methodology. Methodology is the selection of research objects, the questions directed toward them, the design of a study, and the implications that our objectives have for carrying out research. Finally, at the most concrete and practical level we find research methods. In this sense research methods are the ways we go about collecting and analyzing data, and then we draw conclusions from these processes. Schematically, the entire research framework can be represented as follows (after Gomez and Jones 2010):

Ontology and epistemology
↓↑
Paradigms
↓↑
Research methodology
↓↑
Research methods
↓
Conclusions and recommendations

Both directional arrows indicate that theory needs to be responsive, constantly amended and reworked, in light of the surprises and contradictions that emerge in concrete research activities.

Therefore, the entire framework of research is a type of intermediate theory that attempts to connect to all aspects of inquiry. However, the domain of inquiry may vary from case to case, especially in remote sensing. With the change of domain

Table 1.2 Definition and description of the various types of scientific research by UNESCO

Types of research	From the standpoint of the investigator		From the standpoint of the results obtained or expected	
	Motive of the investigator	Degree of freedom of the director of research	Prospective applicability of results	Scientific significance of results
Fundamental Research				
Pure research (free fundamental research)	Research directed toward fuller understanding of nature and discovery of new fields of investigation, with no practical purpose in mind	Choice of field, program and method of work	Delay of practical application unpredictable	Results affect a broad area of science and often have a penetrating and far-reaching effect
Oriented fundamental research (it has two types: field centred research and background research)	*Field centred research* (Exploiting new fields of investigation)	Choice of program and method of work	Delay of practical application is generally long	Results affect a well determined field of science and have a general character
	Research focused on a specific theme, generally connected with a natural phenomena of broad scope, and often directed towards a well-defined objective			
	Background research Research directed towards increased accuracy of scientific knowledge in a particular field by gathering essential data, observations and measurements	Choice of method (and sometimes program) of work	Delay of practical application depends essentially upon field of research	Results are of empirical character and provide the necessary basis facts for the advancement of pure and applied sciences
Applied Research	Research directed towards a specific practical aim, to serve human needs	Choice of method (and exceptionally program) of work	Delay of practical application is generally short	Results generally affect a limited area and have a specialized character
Development Work	Systematic use of the results of applied research and of empirical knowledge directed toward the production and use of new materials, devices, systems and methods for different application areas	Field and program of work laid down by sponsor (sometimes experimental design of research)	Practical application generally immediate	Results affect a very limited area and have a narrowly specialized character

the form of inquiry will essentially vary. Further, there may be disagreements among the outcomes from an inquiry. That means, although the theoretical framework of research seems simple and constant, its atoms may be chaotic. A good researcher, perhaps, can bring this chaos in harmony.

References

Alavipanah SK, Ghazanfari K, Khakbaz B (2010) Remote sensing and image understanding as reflected in poetical literature of Iran. In: Proceedings of remote sensing for science, education, and natural and cultural heritage, 30th symposium of european association of remote sensing laboratories, 31st May–3rd June, UNESCO Headquarters, Paris, France. http://www.earsel.org/symposia/2010-symposium-Paris/Proceedings/EARSeL-Symposium-2010_1-02.pdf. Accessed 29 July 2011

Anderson G (2000) Scientific method. Online lecture note. http://web.archive.org/web/20060217052458/http://pasadena.wr.usgs.gov/office/ganderson/es10/lectures/lecture01/lecture01.html

Bhatta B (2011) Remote sensing and GIS, 2nd edn. Oxford University Press, New Delhi

Busch K (2009) Artistic research and the poetics of knowledge. Art Res 2(2). http://www.artandresearch.org.uk/v2n2/busch.html

Campbell JB (1996) Introduction to remote sensing. Guilford, New York

Deleuze G (1977) Intellectuals and power: a conversation between Michel Foucault and Gilles Deleuze. In: Bouchard DF (ed) Michel Foucault, language, counter-memory, practice: selected essays and interviews. Cornell University Press, Ithaca, pp 205–217

Ennis RH (1987) A taxonomy of critical thinking skills and dispositions. In: Baron JB, Sternberg RJ (eds) Teaching thinking skills: theory and practice. Freeman, New York, pp 9–26

Frield MA, Estes JE, Star JL (1988) Advanced information extraction tools in remote sensing for earth science applications: AI and GIS. AI Appl 2(2–3):17–30

Goldhaber AS, Nieto MM (2010) Photon and graviton mass limits. Rev Mod Phys 82(1):939–979. doi:10.1103/RevModPhys.82.939

Gomez B, Jones JP III (eds) (2010) Research methods in geography: a critical introduction. Wiley-Blackwell, West Sussex

Hoffman RR, Markman AB (eds) (2001) Interpreting remote sensing imagery: human factors. CRC press, Boca Raton

Jarrard RD (2001) Scientific methods. (Online book) http://emotionalcompetency.com/sci/booktoc.html

Jensen JR (2006) Remote sensing of the environment: an earth resource perspective, 2nd edn. Prentice Hall, Upper Saddle River, NJ

Kosso P (2011) A summary of scientific method. Springer, Heidelberg

Lillesand TM, Keifer RW, Chipman JW (2007) Remote sensing and image interpretation, 6th edn. Wiley, New York

Newton I (1999) Philosophiae Naturalis Principia Mathematica (3rd edn). University of California Press, California (trans: Cohen IB, Whitman A)

Schafersman SD (1997) An introduction to science: scientific thinking and the scientific method. (Online lecture note) http://www.geo.sunysb.edu/esp/files/scientific-method.html

Ziemski S (2005) Two types of scientific research. J Gen Philos Sci 10(2):338–342

Chapter 2
Research Framework

Abstract Although this book is focused on research methods, the entire research framework is also necessary to be addressed in context. As stated earlier in Chap. 1, research methods involve some fundamental theoretical questions. These questions are philosophical and concern to the ontology and epistemology. Such philosophical concerns tend to get stored out into distinct paradigms that a researcher can follow. Nested within the theoretical coordinates of paradigms, a set of decisions one has to make about the methodology. Finally, at the most concrete and practical level we find research methods. This chapter is aimed to discuss the entire research framework in the context of remote sensing. Once we gain some ideas on overall research framework, we can proceed further to research methods in the subsequent chapters.

Keywords Remote sensing • Research • Philosophy • Ontology • Image mining • Epistemology • Paradigm • Inductive • Deductive • Technological • Logic • Method • Methodology

2.1 Ontology

Ontology is the largest branch of metaphysics in philosophy, and traditionally deals with questions of existence or reality. Ontology as a branch of philosophy is the science of what is, of the kinds and structures of objects, properties, events, processes and relations in every area of reality. Ontology is often used by philosophers as a synonym for metaphysics—a term which was used by early students of Aristotle to refer to what Aristotle himself called first philosophy. Sometimes ontology is used in a broader sense, to refer to the study of what might exist, where metaphysics is used for the study of which of the various alternative possibilities is true of reality (Smith 1999). Ontology thus provides the basis for exchange of information, and is a fundamental pre-requisite to description and explanation, in science and elsewhere.

In simple words, ontology seeks the classification of entities. Typically, philosophical ontologists produce theories that are very much like scientific theories, but of a far more general nature. Ontology is both a branch of philosophy and a fast-growing component of computer science concerned with the development of formal representations of the entities and relations existing in a variety of

B. Bhatta, *Research Methods in Remote Sensing*, SpringerBriefs in Earth Sciences,   21
DOI: 10.1007/978-94-007-6594-8_2, © The Author(s) 2013

application domains. Ontology has been shown to have considerable potential on the level of both pure research and applications. It provides foundations for diverse technologies in areas such as information integration, natural language processing, data annotation, and the construction of intelligent computer systems.

Recently, the term ontology has been used by information scientists to refer to canonical descriptions of knowledge domains, or associated classificatory theories. In this sense, ontology is "a neutral and computationally tractable description or theory of a given domain which can be accepted and reused by all information gatherers in that domain" (Smith 1999).

Often it is said that remote sensing can provide the 'true' representation of the earth's surface. This statement is never true. Remote sensing provides an impression of the earth-surface features in pictorial format. Pictures are not the real truth; e.g., picture of a flower and the flower itself are not same. Therefore, ontology of remote sensing is primarily the inquiry about the existence or reality through the images. This inquiry is directed towards understanding and defining earth-surface features, spatial relations, processes, their categories, and so on. It would include not only the basic data models, concepts, and representations or classifications of earth-surface features, but also the ontological principles.

2.1.1 Objects and Fields

Remote sensing requires the classification (either object-based or field-based) of earth-surface features. The most widely accepted conceptual data model for spatial information considers that the geographic reality is represented as either fully definable entities (objects) or continuous spatial variations (fields). Objects are with discrete boundaries represented by geometric features; e.g., Roads, buildings, water bodies, etc. Fields are continuous phenomena such as elevation, temperature, and soil chemistry; they exist everywhere. These statements are very simple to the geospatial community. Unfortunately, remote sensing cannot handle them with that much of simplicity.

If we consider individual bands of remote sensing images, they are two-dimensional functions, arising from the sampled response of a region of the earth to an external energy source (the sun or a radar beam) as measured by a passive or active sensor, respectively. In case of thermal and passive microwave remote sensing, the earth itself is the source of energy. Whatever the source of energy or the techniques involved in capturing that energy, remote sensing images are always continuous, not discrete. Further the properties of each sensor (i.e., the number of bands as well as the spectral, temporal, radiometric, and spatial resolutions) are the results of a compromise between the needs of various research communities and the availability of sensor technology. The continuous variation of the spectral response of the land-cover, which is the specific phenomena captured by the pixel values, often misses what a domain scientist considers as relevant. These measurements are merely components of the more complex information content of an image. Most image classification techniques do not rely explicitly on the conversion between digital counts (pixel values) and the actual energy captured by the sensor, but they use the digital counts to extract features. As a consequence, viewing images as fields of values of reflected energy is insufficient for their ontological characterization (Câmara et al. 2001). The limitations of the field perspective to the ontology of images have led some researchers to view a remotely sensed image as a container of an implicit set of objects, which are extracted by manual or semi-automated analysis procedures. But important to realize, the real-world boundaries exist independent of human cognitive acts. Further, measurement of reflected/emitted light and the identification of objects via manual or semi-automated analysis are independent, not controlled or operated by one another. For example, identification of objects does not consider the real atmospheric contributions that might have recorded by the sensor. Furthermore, the pixel is a generalized representation of reality; mostly they are mixed in nature—*mixel* (mixed pixel). A pixel cannot be divided further to represent an object smaller than the pixel size.

2.1.2 Classes

Class refers to a group of features identical or similar types that have taxonomic significance. Classification is an abstraction mechanism that maps individuals into a common class. Classes are often interrelated by a generalization processes, capturing different levels of detail about the same individuals. For example, deciduous and evergreen are two classes that can be generalized as forest. In this case, ontology describes a hierarchy of concepts created by a generalization process. Although the object perspective captures a fundamental component of the ontology of images and forms a basis for a large set of image classification techniques, it is still incomplete. In many cases, there is no corresponding object in the world, since we deal with purely physical phenomena; e.g., ground elevation—that may change in every pixel. Actually, object-based classification is an attempt for geospatial segmentation of earth-surface features based on the pixels and their digital values. This helps us to understand the physical earth-surface as a distinctly segmented space. This concept overlooks the existence of 'mixel' (mixed pixel).

Field-based sub-pixel classification, on the other hand, considers the geographic space as continuous phenomena. It tries to quantify the amount of a feature or material within a pixel. The entire pixel or a part of it may be occupied by a material. Therefore, it calculates the percentage or amount of that material within a pixel. This concept handles the problem of mixel well, but ignores the identification of other surface materials. Therefore, there is no spatial segmentation among the features, and thus, no objects in existence.

The preceding discussion shows that neither of the classification approaches is sufficient by itself to support the full process of knowledge representation for remotely sensed images. The underlying reason is that images have a dual nature: earth surface features within them can be interpreted as fields as well as objects; although, they are fields at the measurement level.

2.1.3 Relations

Remote sensing ontologies are different from many of other ontologies in that it embraces spatial relations that play a major role in the geospatial domain. Earth-surface features can be connected or contiguous, scattered or separated, closed or open, near or far, and so on. These relations not only exist in geographic form but also in concepts. For example, two large forests may be connected by a narrow forested passage (geographic form), and these two forests may be connected in the sense that both are of same species (conceptual form). Therefore, 'part and whole' relations and other spatial relations are needed to be described in remote sensing ontologies.

2.1.4 Functions

Functions are mappings or transformations applied on remotely sensed images, and can be of many types. Basic image processing functions, such as filtering, principal component transformations, resampling, or interpolation, are familiar to and used by almost all researchers working with remote sensing data. Other generic functions exist in the statistical analyses of spatial domain, e.g., spatial statistics or spatial metrics. Of more significance here are functions that map the state of earth's surface can be applied as a process because earth-surface features are not static—they are continuously changing with the change of time. For example, spatial metrics can determine the built-up pattern; but how the built-up pattern changes with time is essentially a process that requires temporal consideration. Our concept of remote sensing ontology thus includes notions both of form (shape/size/pattern) and process. The process does not refer only to historical process but simulations for the future as well.

Câmara et al. (2001) argued that "a geographic landscape is an ever-changing scenario, and the process of data capture by remote sensing satellites implies that an image is a measurement that captures snapshots of change trajectories. Therefore, the focus of the ontological characterization of images should be on the *search for changes* instead of the *search for content*. The emphasis of such ontologies should not be placed on simple object matching and identification procedures, but on capturing dynamics over a finite landscape."

2.1.5 Image Ontology

Câmara et al. (2001) proposed a multi-level ontology for images, based on the concept of *action-driven ontologies* for GIS (Câmara et al. 2000). The authors considered that remote sensing images are ontological instruments to capture landscape dynamics. The proposal takes into account that images have a particular, distinct description independent of the domain ontology a scientist would employ to extract information. The ontology domain for images has three interrelated components:

1. *Physical ontology*—describes the physical process of the image creation, focusing the knowledge about the relation between the reflected energy by terrain surface and measures obtained by the sensor.
2. *Structural Ontology*—contemplates geometric, functional and descriptive structures that can be extracted using techniques for feature extraction, segmentation, classification, and so on.
3. *Method Ontology*—it is composed of a set of algorithms (that perform transformations from the physical level to the structural level) and data structures that represent reusable knowledge in the form of image processing techniques (filtering, smoothing, and others).

The algorithms that are part of the *method ontology*, perform transformations from the physical level to the structural level, a process than can be called *structural identification*. When applied to an image (or a set of images), this process results in a set of *structures* strongly related to the measurement device properties and its interaction with the physical landscape. These structures may be geometric (e.g., regions extracted by a segmentation procedure, i.e., per-pixel classification) or functional (e.g., normalized differential vegetation index or sub-pixel classification) (Câmara et al. 2001).

2.1.6 *Image Mining and Image Ontology*

Image mining deals with extraction of implicit knowledge, image data relationship or other patterns not explicitly stored in images and uses ideas from computer vision, image processing, image retrieval, data mining, machine learning, databases and artificial intelligence. The fundamental challenge in image mining is to determine how low-level pixel representation contained in an image or an image sequence can be effectively and efficiently processed to identify high-level spatial objects and relationships. Typical image mining process involves pre-processing, transformations and feature extraction, mining (to discover significant patterns out of extracted features), evaluation and interpretation and obtaining the final knowledge. Various techniques from existing domains are also applied to image mining and include object recognition, learning, clustering and classification, just to name a few (Zhang et al. 2002).

Extracting information from images remains a complex and tedious process; sometimes inferior in respect of our needs. Our capacity to build sophisticated remote sensing sensors is not matched by our means of producing information from these data sources. Currently, most image processing techniques are designed to operate on a single image, and we have few algorithms and techniques for handling multi-temporal images. This situation has lead to a 'knowledge gap' in the process of deriving information from images and digital maps (MacDonald 2002). This 'knowledge gap' has arisen because there are currently few techniques for image mining and information extraction in large image datasets; thus we are failing to exploit our large remote sensing archives. Image ontology facilitates the deployment of the concept for various classes of models for the information extraction from the remote sensing imagery. Silva and Câmara (2004) presented the architecture of ontology based image mining. Durand et al. (2007) also explained the ontology-based object recognition for remote sensing image interpretation.

2.2 Epistemology

Epistemology is the study of knowledge. It deals with the nature of knowledge, how do we know things, what do we know, why we know, is what we know true, and what are the limits of knowledge. It is true that we know several things. But what is the nature of what we know? One should realize that research is only one of several ways of 'knowing'. Epistemologists generally recognize at least four different sources of knowledge:

1. *Intuitive knowledge* takes forms such as belief, faith, intuition, etc. It is based on feelings rather than 'facts'.
2. *Authoritative knowledge* is based on information received from people, books, journals, newspapers etc. Its strength depends on the strength of these sources.
3. *Logical knowledge* is arrived at by reasoning from 'point A' (which is generally accepted) to 'point B' (the new knowledge).
4. *Empirical knowledge* is based on demonstrable, objective facts (which are determined through observation and/or experimentation).

General research often makes use of all four of these ways of knowing: intuitive (when coming up with an initial idea for research), authoritative (when reviewing the professional literature), logical (when reasoning from findings to conclusions), empirical (when engaging in procedures that lead to these findings).

However, in case of remote sensing, these four sources of knowledge can be viewed from a different perspective in addition to the ways exampled above. In the field of remote sensing, understanding the epistemological aspects of the interpretation and information extraction is crucial. The real environment is spatio-temporally very complex; whereas, remote sensing images can be regarded as a snapshot (obviously simplified) of this complex environment. Spatial, spectral and contextual evidence can be implicitly or explicitly linked to surface phenomenon. Inference is a crucial methodological element of the transformation from these facts/evidences to knowledge (Miller and Han 2001; Gahegan 2001). Inference requires all four ways of knowing—intuitive, authoritative, logical, and empirical. Further, remote sensing has a huge data archive for the past. As a result, many empirical studies can be done within this discipline. Multi-temporal datasets can form the basis of our empirical knowledge in such cases.

2.3 Research Paradigm

Having stressed the importance of reflecting on the nature of the world, and how we come to gain knowledge of the world—a twin process that should run throughout the research process—we are now in a position to discuss in more detail how ontology and epistemology find their place within different paradigms in our research. "A paradigm is a world view, a general perspective, a way of breaking down the complexity of the real world" (Patton 1990). It is the underlying assumptions and intellectual structure upon which the research and development in a field of inquiry is based. "A paradigm may be viewed as a set of basic beliefs ... that deals with ultimates or first principles. It represents a worldview that defines for its holder, the nature of the 'world', ... and the range of possible relationships to that world and its parts" (Guba and Lincoln 1994). Henning et al. (2004) define the paradigm as "a theory or hypothesis", a paradigm is rather a framework within which theories are built, that fundamentally influences how we see the world, determines our perspective, and shapes our understanding of how things are connected. Holding a particular world view influences our personal behaviour, our professional practice, and ultimately the position we take with regard to the subject of our research. Paradigm is an interpretative framework, which is guided by a set of beliefs and feelings about the world and how it should be understood and studied. The word paradigm has been used in science to describe distinct concepts. The word has come to refer very often now to a thought pattern in any scientific discipline or other epistemological context. According to Kuhn (1996), there can be no science without a paradigm. There must be this stable, authoritative body of commitments for scientists to communicate and do more than start from scratch each morning. Thus, paradigm refers here to a body of literature that shares fundamental assumptions about what the world is like and how we should research it, but also, and more specifically, about what the key objects of analysis should be, and the role of remote sensing within the real world.

One should remember that within particular projects both methodology and methods can overlap and, further, each paradigm contains within it some often causes debates. Kuhn (1996) wrote that "successive transition from one paradigm to another via revolution is the usual developmental pattern of mature science". However, a scientific paradigm changes on very rare occasions. Paradigm-shifts tend to be most dramatic in sciences that appear to be stable and mature. Perhaps the greatest barrier to a paradigm shift, in some cases, is the reality of paradigm paralysis: the inability or refusal to see beyond the current models of thinking. Considering the nature of paradigm, the change cannot be piecemeal, partial, or gradual. A paradigm shift must be wholesale and abrupt. It must be a scientific revolution. Kuhn's (1996) motivation for saying this is based on his survey of historical cases, but it could just as well be argued in principle, based on the essentially holistic nature of paradigms. The influence of a paradigm is permanent in that it establishes the appropriate language of the science, the background knowledge against which new ideas are compared, and even the standards by which new experimental results are judged. All aspects of a science are thus paradigm-dependent, including any guidelines for judging the acceptability of a paradigm. In other words, paradigms underwrite their own acceptability. Change from one paradigm to another, according to Kuhn (1996), is not governed by any rational, methodical standards. Since each paradigm judges itself by its own standards, there are no external, paradigm-independent guidelines for determining that one paradigm is better, more likely to be true, than another. During a revolution, a paradigm shift, the change cannot be measured as progress, since no standard of measurement is preserved from one paradigm to the next. We of course judge old paradigms as less accurate than our own. But this, according to Kuhn (1996), is an inappropriate application of the standards of one paradigm to the evaluation of another. Given the holistic, pervasive nature of a paradigm, each must be judged by its own standards.

Spatial science (and thus remote sensing) rests on the foundational pillars of objectivity and generality in searching for orderly causal processes; it also adopts what is said to be a realist approach to representation (Abler et al. 1971). Ontologically speaking, as stated by Shaw et al. (2010), spatial scientists maintain a strict divide between space and time, and between space and society. In turn, both space and time are measurable, insofar they can be divided into increments, and these increments can be empirically assessed. Within this paradigm, remote sensing can be taken to be synonymous with the term 'spatial', such that remote sensing research becomes a matter of asking: (1) how do objects and practices vary across the earth's surface; (2) why do these variations take the spatial forms that they do; or simply (3) what present at a spatial location and why. Descriptive accounts of spatial variation (case 1), then, should in turn lead to explanatory ones (case 2). The latter typically rely on the development of qualitatively- or quantitatively-stated conceptual models that set forth hypothetical explanations for the variations of interest. Research questions suggested by the models lead to data collection and analysis (again, these can be qualitative or quantitative). Importantly, this type of research is made possible by assuming that the objects of analysis are discrete; this enables questions that take the form of the following: "how does the spatial distribution of X compare with that of Y"; "how does X cause Y"; and "how do X and Y cause each other".

A useful paradigm for remote sensing analysis is in terms of the degree-of-indirectness with which the measured radiation field and the wanted quantity are linked (Quenzel 1983). Within this framework, analysis models are arranged according to a particular 'model order'. As the order of the model increases, the link between the measured radiation field and the quantity of interest become more remote. The approach is more conceptual than pure physics, but it gives some insight into the remote sensing problem (Quenzel 1983). The framework enables one to conceptualize the level of accuracy achievable for the quantities derived from the remote sensing imageries, because error increases as the link becomes more complex (i.e., as the model order increases).

Examples (from Prenzel 2004) of a first-order analysis are the extraction of surface reflectance from raw radiance values, whereas a second-order analysis would involve the extraction of some type of measured surface characteristic, such as surface temperature or ozone concentration in the atmosphere. Third-order analysis derives several biophysical parameters, for example, land-cover, leaf area index, biomass, or leaf pigment content. Fourth-order analysis derives parameters—for example—fuzzy and thematic measures of land-use, measures of net primary productivity. The difficulty with accurately extracting high-order parameters such as land-use is that they are not closely linked to the physical properties of terrain that are measured by remote sensing sensors.

A second and complementary paradigm of describing remote sensing analytical models is in terms of their 'determinism' (Schott 1997). Generally, these models can be thought of as occurring along a continuum between being deterministic and being empirical. The basic difference between these two concepts is that deterministic processes can be quantitatively described and predicted very accurately (i.e. close to 100 %), whereas empirical processes can only be quantitatively described in terms of levels of confidence. It may be worth mentioning that a well developed deterministic model can be inverted to accurately predict biophysical variables from the original remote sensing data beyond the confines of the image from which the model was developed (Quenzel 1983). Determinism is the view that every event, including human cognition, behaviour, decision, and action, is determined. In contrast, empirical models cannot be used to accurately predict parameters beyond the image data from which they were derived. Perhaps, the main drawback of deterministic models is that they are typically more demanding in terms of data inputs, model establishment, and model validation (Jensen 2005). Whereas, empirical models are often easier to develop and apply, since they are generally less demanding in terms of analysis, model establishment, and validation.

2.4 Methodology

A research methodology defines what the activity of research is, how to proceed, how to measure progress, and what constitutes success. Research methodology is generally a guideline for solving a problem, with specific components such as phases, tasks, methods, techniques and tools (Irny and Rose 2005). Generally speaking, methodology does not describe specific methods despite the attention given to the nature and kinds of processes to be followed in a given procedure or in attaining an objective. In theoretical concept, the development of paradigms satisfies most or all of the criteria for methodology. Nested within the theoretical coordinates of paradigms, a set of decisions one has to make about the methodology.

Research methodologies are always situated within larger theories of the world. Many of the most creative aspects of the research process involve questions that translate those theories into more precise research objectives, questions and tasks. To these ends, it can be helpful to formulate a research methodology in terms of a series of questions, the most basic of which include the following (after Gomez and Jones 2010):

1. What objects or events should I select to analyze?
2. How should I theorize their domain of operation?
3. How should I theorize their relationships to other objects and events?
4. What research questions are appropriate for explaining or understanding them?
5. What data are to be collected for answering those questions?
6. How should I collect the required data?
7. What procedures should be used to analyze that data (considering available alternatives)?
8. Are the available models/procedures sufficient to analyze the data?
9. Does it require developing new models/procedures?
10. What safeguards should we rely upon to ensure the validity and reliability of our account?
11. What are the grounds for evaluating competing accounts?
12. How are my findings influenced by the 'theoretical priors' I brought to the research?
13. What is the purpose of the research (e.g., the production of scientific/technologic knowledge, saving the earth, transforming society, something else)?
14. What ethical safeguards have been followed or needed to be addressed?

Having posed these and other questions about the world and the intended research activity, a number of obvious connections to specific research methods will emerge.

The first step in remote sensing, as in any scientific study, is the definition of a problem. Due to its multidisciplinary nature, the problems that remote sensing can be applied to are numerous and diverse. In spite of this, the approaches to remote sensing can be categorized as being either scientific or technological in nature. The distinction is primarily a function of the motive behind solving the problem. The methodology that is subsequently applied to the problem is usually dependent upon the origin of the problem. Scientific approach is this process of exploring data to gain an intimate knowledge of it and, hopefully, insight into it. However, the primary goal of remote sensing is not only the pursuit of knowledge, but also the application of any knowledge gained.

There are three basic types of logic that can be applied to a problem; inductive, deductive, and technologic. Scientific approaches use both inductive logic and deductive logic methodologies, while a technological approach uses a technologic logic methodology. In the scientific approaches the motivation is curiosity, the goal is knowledge and the methodology is often induction to derive theory and the deduction to verify the theory. In the technological approaches the motivation is human need, the goal is the application of knowledge and the methodology is design. The inductive approach uses the data to generate ideas. The deductive method starts with an idea or theoretical framework and uses the data to verify the idea in order to prove or disprove the idea (Holloway 1997). Often a combination of both approaches is used.

2.4.1 Inductive Logic

Inductive logic could be described as learning logic. The inductive methodology seeks to form tenable theories by making observations of phenomena, classifying these observations, and making generalizations that form the basis of theories. Inductive generalization is a very common-sensible process. If we observe that something A_1 has a property B, and that A_2 is B, and that A_3 is B, and so on, we eventually conclude that all A's are B (Kosso 2011). This is bottom-up, outside-in reasoning. The conclusion has come from nowhere other than observations of nature, without fabrication. It is understood that the conclusion is only probably true, but that is no surprise, given our respect for the fact that science does not deliver certainty.

Inductive reasoning works moving from specific observations to broader generalizations and theories. Informally, sometimes this is called 'bottom-up' approach (Fig. 2.1). Conclusion is likely based on premises. It certainly involves a degree of uncertainty. Most people use inductive logic every day. For example, a person slips and falls on water which has spilled on the bathroom floor. He would make the observation that when the tile in the bathroom is wet, there is subsequent loss of traction. This observation can then be generalized to a theory that 'all tiles, when wet, provide less traction than when dry' (Jensen 2006).

Now think about the following statement, which may also be concluded by examining numerous colour-infrared images using same concept: 'healthy vegetation appears red on infrared colour-composite'. Before such a theory can be considered, three conditions must be satisfied (Jensen 2006): (1) the number of observations must be large, (2) the observation must be repeated under a wide range of conditions (variations), and (3) no accepted observation should ever conflict with the derived theory. The

Fig. 2.1 Inductive logic (*left*) versus deductive logic (*right*)

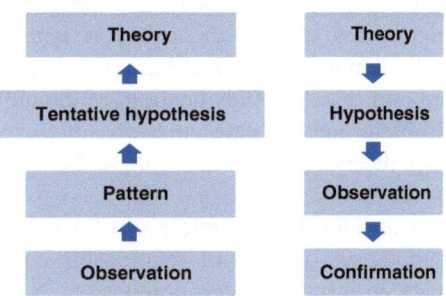

most serious limitation of inductive logic is that no number of apparently confirming observations can ever show that a theory is completely true.

Inductive logic is at the centre of remote sensing when the focus is image interpretation. Like our everyday learning experiences, a researcher using this logic observes facts about remotely sensed data and seeks to form general theories or principles that can be applied to other remotely sensed data (Curran 1987). Theories formed from this inductive approach are often fed directly into a deductive methodology where hypotheses are developed for testing the theories.

2.4.2 Deductive Logic

The deductive logic lays an emphasis not on observation, but on the formulation of theories and the subsequent testing of hypotheses. Once a problem is identified, a researcher conjectures a theory to solve it. To determine the validity of any such theory, hypotheses are developed and tested. Deductive methodology works from more general to the more specific. Sometimes this is informally called 'top-down' approach (Fig. 2.1). Conclusion follows logically from premises (available facts). Induction is usually described as moving from the specific to the general, while deduction begins with the general and ends with the specific.

The hypotheses are at the core of the deductive logic. Because of their importance, great care should be taken to formulate a hypothesis that is appropriate to the problem at hand. Two of the most common types of hypotheses are the factual and the inferential. A factual hypothesis clearly states a position that can be either verified or falsified. It is possible to verify this hypothesis as either truth or falsehood. An inferential hypothesis is one which can be only be falsified. Observations that fail to disprove the hypothesis do not necessarily prove its truthfulness. However, a failure to disprove the hypothesis generally results in the acceptance of the theory being tested with the knowledge that future observations may later reverse that decision (Jensen 2006).

In the deductive logic a null hypothesis is formed, for example, if the hypothesis is 'man has two eyes', the null hypothesis can be formed as 'man does not have two eyes'. The null hypothesis is then tested at specific statistical confidence levels (e.g., 0.05 or 0.001). If the observations are such that the null hypothesis can be rejected, then the theory can be considered acceptable in the guarded sense that there is not an empirical basis for doubting its validity. If the observations do not support rejection (falsification) of the null hypothesis, then we must go back to our problem and evaluate other possible explanations that might lead to a falsifiable hypothesis. A failure to disprove the hypothesis generally results in the acceptance of the theory.

Consider the example given for inductive logic and develop the following null hypothesis: 'there is no relationship between healthy vegetation and a red appearance on infrared colour composite'. Now test the null hypothesis with the ground and if the healthy vegetation at 99 of the 100 sites have a red appearance in the infrared colour composite, then it would be possible for us statistically to reject (falsify) the null hypothesis. We could then state that 'it is confirmed that healthy vegetation appears red on infrared colour composite'. It should be mentioned that in many scientific disciplines 100 % confidence is required for the acceptance of the theory. However, in case of remote sensing, in many instances, 100 % confidence cannot be achieved because it involves science, technology, as well as art. Further, remote sensing observations (imageries) are simplified representation of the reality.

2.4.3 Which Logic to Follow: Inductive or Deductive?

Some people, Sir Isaac Newton among them, have claimed that inductive generalization from observation to theory is all there is to scientific method (Kosso 2011). Anything else, such as speculative hypothesis, would be an irresponsible first step on a slippery slope to make-believe and mysticism. But inductive generalization cannot be the whole story about scientific method; let us see the specific reasons why.

First of all, pure induction, with only observations as premises, could never imply a statement about something unobserved or unobservable. How did we ever come up with ideas about germs or atoms? How do we diagnose a disease or identify a season? It could not have been simply by generalizing on what has been observed, since none of these things has been observed. If the goal of science was simply to catalogue empirical generalizations like all metals conduct electricity and all grasses are green, then pure induction might do the trick. But science routinely does more than that. It offers explanations for how and why metals conduct electricity and grasses reflect green light. This is the value of science, getting beyond the merely observable, and pure induction does not suffice as the way to do this.

There is a second, and more fundamental, reason why induction cannot be the whole story, or even the most important part of the story, about scientific method. Pure induction presupposes pure observation, an uncontaminated flow of information from outside-in. This is simply impossible. In life as in science, perception is influenced by ideas. Scientific observations are influenced by scientific theories, so the order of events cannot be strictly observation then theory. It is an important insight into scientific method to show not only that theory influences observation, but exactly how this influence comes about. Kosso (2011) explained four reasons why there can be no theory-neutral observations in science. First, it is impossible to observe everything. Selecting what to observe and what to ignore cannot be haphazard. Science is not simply a catalogue of observations; it is inferences from relevant observations. For example, land-use information cannot be obtained through simple observation of remote sensing imagery; simple observation can provide simplified land-cover information. Land-use information must be inferred from the observations about land-cover. Some basic idea of what is relevant to what is needed in selecting what observations to make and record. Isaac Newton understood the relevance between a falling apple and the orbit of the moon, but no one else did. No one else had the theoretical predisposition to know what to make of the apple and to know that it should be recorded as an important observation. Without some theory in mind, the falling apple would have been unremarkable and unrecorded. Secondly, it is impossible to note and describe every detail of the observations that are made. Selections have to be made, this time regarding the relevant aspects of the observations. And again, the selections are not random or haphazard, but informed by some existing understanding of the situation. Newton understood that the colour or shape or size of the apple is not relevant rather it is the mass of the apple. A third reason for some theoretical imprint on the information from observation is the requirement that scientific observations be careful and reliable; because observations can be good or bad. Observing conditions must be proper; because it may highly influence the observation (e.g., atmospheric condition while capturing the remote sensing imagery). Relevant conditions must be controlled. If machines are used, they must be working properly, and so on. Accounting for reliability will call on a theoretical understanding of how the remote sensing sensors work, which conditions are relevant, and what amounts to proper conditions. Scientific observation, unlike casual observation out on the street, is accountable; it is always open to challenges to its accuracy. Meeting the challenge calls on background knowledge. Finally, scientific observations must eventually be rendered in theoretical language, and this certainly presupposes a theory in place.

Scientific claims must be based on observable evidence, and we have seen that the observations that come before a theory are insufficient basis for claiming the theory is true. So there must be an important role for evidence that comes after the theory is proposed, evidence used to test it. If a statement is about something that is itself observable, then the empirical testing can be direct. We just have a look to see if it is true. For example, the statement, "grass is green", is subject to direct empirical testing. But science is most interesting and most useful to us when it is describing unobservable things like atoms, germs, black holes, gravity, the process of evolution as it happened in the past, and so on. This is where we will find explanations, and not mere summaries, of what happens in nature. Theories, that is, claims about unobservable things, are not amenable to direct empirical testing, since we cannot just have a look to see if they are true. These claims are nonetheless accountable to empirical testing that is indirect. The nature of this indirect evidence, and the logical relation between evidence and theory, are the crux of scientific method (Kosso 2011). This is where we have to be the most careful and explicit. Statements about unobservable things can be tested by their observable implications. In other words, to test the truth of a statement x, we reason that "if x is true, then we will observe y". y is an observable implication from x, and it is by observing y that x is indirectly confirmed. If we look for y but do not see it, then x is indirectly disconfirmed (falsified). y is the evidence for (or against) x. Here is a simple example: "if malaria (x) then there should be fever (y)". This kind of statement, if x then y, is the central premise of indirect empirical testing. Since it is a case of deducing the prediction y from the hypothesis x, any test that involves an if-then statement like this is called *hypothetico-deductive* testing. Consider another example from remote sensing, "if buildings are too large on the imagery (x) then they are being used for commercial or industrial purpose (y)".

However, this logic often commits fallacy. Any doctor knows that if someone has malaria they will have a fever. But the evidence of a fever does not warrant the conclusion that a patient has malaria. If a building is being used for commercial/industrial purpose it does not always be too large. But now suppose the prediction had come out to be false? That is, suppose the hypothesis is "if malaria (x) then there should not be fever (not y)". Now, the observation shows that there is always fever when malaria. Therefore, the conclusion is that the hypothesis is false. This argument is valid. There is no way that the hypothesis could be true and the conclusion false. The conclusion in this form of argument follows with absolute certainty. Applying this to the case of diagnosing malaria, a patient who does not have a fever cannot possibly have malaria.

It seems as if disconfirmation of a hypothesis can be done with a single test, if the prediction is false. Disproof of a hypothesis, falsification, appears to be decisive in a way that proof is not. This apparent disparity between falsification and confirmation has led many people to claim that the essence of scientific method is falsification. Scientists do not prove theories, according to this account; they try to disprove them. Theories that survive repeated attempts at falsification are the ones to believe. But this is wrong. Consider the second example, "if buildings are too large on the imagery (x) then they are *not* being used for commercial or industrial purpose (not y)". Now, if we find a case that the building is too large and it is being used for commercial/industrial purpose, we cannot immediately disprove the null-hypothesis and conclude that "if no commercial/industrial use then buildings are not too large" (larger buildings are also used for other purposes, e.g., multi-storied large residential complex). Remember the falsification of null-hypothesis had worked for malaria (if not fever then no malaria). This suggests that falsification of null hypothesis is not reliable in every case. Disconfirmation sometimes seems so easy and so definitive only because we have ignored many of the important details. In particular, we have ignored the theoretical details of how the prediction was deduced in the first place, and the practical details of how the experiment was done.

So where does this leave us in our understanding of scientific method? Empirical testing (inductive logic) is never decisive. A false hypothesis (deductive logic) can make a true prediction, just as a true hypothesis can make a false prediction. However, this does not make scientific testing worthless. "A true prediction is still some indication that the hypothesis might be true, and a false prediction forces us to rethink some aspect of the situation, the hypothesis or the auxiliaries" (Kosso 2011). But this is supposed to be the scientific method, and there must be something more methodical than is suggested by all the vague language of 'some' and 'might'. The scientific method is based on evidence and logic, but the details of empirical testing show that evidence and logic alone do not settle the issue of which theories are likely to be true and which false. So what else is involved?

Scientific method must involve a broad view of lots of different ideas. A theory is judged by its relation to many different observations and many other theories. Scientific knowledge must be a coherent network of statements, both empirical and theoretical. Science itself is neither bottom-up nor top-down. It is meet-in-the-middle, between theoretical and empirical information. No piecemeal or isolated view of single theories and single experiments can do justice to the scientific method. Kosso (2011) explained this concept elaborately. The good reason to believe a scientific claim is a matter of overall fit into a coherent network of other claims, theoretical and empirical. This means that no one can responsibly judge the plausibility of a scientific claim quickly or in isolation from its broad context. There is no short course to scientific expertise, no short cut to making informed decisions. Scientific judgment is informed judgment, and that requires knowing a lot of the context.

A researcher needs to realize that the structure of everyday knowledge differs from the structure of scientific knowledge. In the everyday, we all share a common network of basic background beliefs. We all have the expertise required to put ideas in context, and we do it implicitly and without really thinking about it. So we can all make informed judgments about the everyday ideas. We are our own well-qualified experts in life, but not in science.

The concept of coherence in a network of ideas is playing an important role in this account of scientific method, and it deserves as much precision as possible. The most basic requirement of coherence is logical consistency. A network of scientific knowledge cannot tolerate contradiction. This is not to say that there are no contradictions lingering in the sciences, but where they are identified they must be addressed. Contradiction cannot be ignored.

Scientific claims must not only be consistent, they must be cooperative. This is less precise than logical consistency, but it requires not just compatibility in the network of ideas, but connections among the ideas. Theoretical claims explain observations, and sometimes they explain other theoretical claims. One theory participates in the role of auxiliary in accounting for the evidence of another theory, and so on. There is a variety of kinds of links between scientific ideas. And building such an interrelated, coherent web of claims is a challenge and an accomplishment. Scientific publications often have more than one author. Many authors can bring many areas of expertise and thereby link many ideas together.

It is worth repeating the big idea at this point, because it is often over-looked, even though it is an essential aspect of scientific credibility. Scientific method must consider the broader network of ideas. Furthermore, science is not a process of settling one issue and moving on to the next, accumulating truths. There is an ongoing back-and-forth among ideas, between theory and observation, and between theory and theory. "Entries in this dynamic web of knowledge are never written in pen. Science is recorded in pencil. The classification of being hypothetical wears off slowly. It wears off as the theory is linked to more and more observations and to more and more other theories. As the hypothesis comes into equilibrium within the network of scientific knowledge, there is more and more good reason to believe it is true. This is the scientific method" (Kosso 2011).

2.4.4 Technological Approach

A technological approach differs from both the inductive and deductive in both its origin and its goal. The basis of this approach is human need rather than scientific inquiry. The goal is the rectification of that need rather than simply an increase in knowledge. The focus of a technological method is the design of coherent plan which successfully blends 'inputs from science, economics, aesthetics, law, logistics and other areas of human endeavour' (Curran 1987). Once a plan of action has been designed it is implemented without a formal hypothesis being stated. Some scientists extract new thematic information directly from remotely sensed imagery without ever using inductive or deductive logic. They are just interested in extracting information from the imagery using appropriate methods and technology. Although this approach is not always acceptable, it is common in applied remote sensing. This approach can also generate new knowledge.

2.5 Research Methods

Research method refers to specific procedures used to gather and analyze the data. Research methods and research methodology are two terms that are often used interchangeably. Strictly speaking they are not same. One of the primary differences between them is that research methods are the methods by which one conducts research into a subject or a topic; whereas, research methodology explains the methods by which one may proceed with their research. Research methods involve conduct of experiments, tests, surveys and the like. On the other hand research methodology involves the learning of the various techniques that can be used in the conduct of research and in the conduct of tests, experiments, surveys and critical studies. In short it can be said that research methods aim at finding solutions to research problems. On the other hand research methodology aims at the employment of the correct procedures to find out solutions. It is thus interesting to note that research methodology paves the way for research methods to be conducted properly. Research methodology is the beginning whereas research methods are the end of any scientific research.

According to the research paradigm in remote sensing, research methods can be of deterministic and empirical. Empirical methods include: (1) image subtraction, (2) classification, (3) neural network approaches, (4) image index (ratioing), amongst others (refer Bhatta 2011). Empirical approaches have been applied in a variety of mapping contexts; for example, forestry (Verbyla and Richardson 1996), urban–rural fringe (Gao and Skillcorn 1998), agriculture (Salem et al. 1995), land-cover (Langford and Bell 1997), and land-use (King 1994). Deterministic-oriented analyses have been used to characterize: (1) chlorophyll content (Zarco-Tejada et al. 2001), (2) leaf area index (Chen et al. 1997), (3) chlorophyll fluorescence (Zarco-Tejada 2000), (4) nitrogen content (Johnson 2001), and (5) net primary productivity (Gower et al. 1999). Empirical and deterministic models can also be mixed that are neither strongly empirical nor deterministic. Mixed models are confluence of empirical and deterministic approaches. Some examples are: (1) linear spectral unmixing (e.g. estimating biophysical forest parameters; Peddle et al. 1999), (2) non-linear spectral unmixing (e.g. mapping dense vegetation; Ray and Murray 1996), and (3) physically based classification (e.g. land-cover mapping; Zarco-Tejada and Miller 1999).

As documented in Sect. 1.1.1 (Chap. 1), remote sensing researches can be conducted in a wide variety of disciplines. Accordingly, research methods (i.e., methods involves in collecting and analyzing the data) will also vary at a wide spectrum. Forthcoming chapters will explain the remote sensing research methods in detail.

2.6 Conclusions and Recommendations

For any research project and any scientific discipline, drawing conclusions is the final and most important part of the research framework. Whatever the research methods were used, the final conclusion is critical, determining success or failure. If an otherwise excellent experiment is summarized by a weak conclusion, the results will not be taken seriously. Success or failure is not a measure of whether a hypothesis is accepted or rejected, because both results still advance scientific knowledge. The key is to establish 'what the results mean'. Shuttleworth (2008) documented following important issues that are to be addressed while concluding a research:

What has been learned: Generally, a researcher summarizes what they believe has been learned from the research, and will try to assess the strength of the hypothesis. In observational research, with no hypothesis, the researcher will analyze the findings, and establish if any valuable new information has been uncovered.

Generating leads for future research: Very few experiments give clear-cut results, and most research uncovers more questions than answers. The researcher can use these to suggest interesting directions for further study. If, for example, the null hypothesis was accepted, there may still have been trends apparent within the results. These could form the basis of further study, or experimental refinement and redesign.

Evaluating limitations of the research: The researcher evaluates any apparent problems with the experiment. This involves critically evaluating any weaknesses and errors in the design, which may have influenced the results. Even strict and 'true experimental' designs have to make compromises, and the researcher must be thorough in pointing these out, justifying the methodology and reasoning. For example, when drawing conclusions, the researcher may think that another causal effect influenced the results, and that this variable was not eliminated during the experimental process. A refined version of the experiment may help to achieve better results, if the new effect is included in the design process. As an example, the researcher might establish that 'carbon dioxide emission alone cannot be responsible for global warming'. They may decide that another effect is contributing, so propose that 'methane may also be a factor in global warming'. A new study would incorporate methane into the research.

Documenting the benefits of the research: This is to evaluate the advantages and benefits of the research. In remote sensing, for example, the results may throw out a new (but more effective) model to analyze the vegetation health, so the advantages are obvious. However, all well constructed research is useful, even if it is just adding to the fount of human knowledge.

Recommendations: The final stage is the researcher's recommendations based upon the results, depending upon the field of study. This area of the research process can be based around the researcher's personal opinion, and also integrates previous studies. For example, findings of a research of forest health analysis may recommend some policies or regulations or protective measures to improve the current health status. These may be the researcher's personal opinions or they can refer any previous study that proved such policies/regulations/protective measures to be useful.

Finally, one has to understand the distinction between the conclusions and recommendations. A conclusion is a summary as well as the implications of the research. To give a simple example: if one finds that urban growth rate in a city is very high and the built-up areas are becoming more dispersed, they would summarize these findings and then say 'this implies that the urban growth in the city is not sustainable'. A recommendation is what the researcher believes to be done for the good, or in future research to clarify the results, as well as where the researcher believes a solution to a problem might be. So, to use the same example again in recommendation: 'urban growth control policies should be imposed to restrict this type of growth' or 'it is recommended that future research should focus the issue on how this unsustainable growth of the city can be restricted'. Remember, recommendations must be logical, not hypothetical or intuitive.

References

Abler R, Adams J, Gould P (1971) Spatial organization: the geographer's view of the world. Prentice-Hall, Englewood Cliffs

Câmara G, Egenhofer M, Fonseca F, Monteiro AMV (2001) What's in an image? In: Spatial information theory: foundations of geographic information science, international conference, COSIT

Câmara GM, Monteiro AMV, Paiva JAC, Souza RCM (2000) Action-driven ontologies of the geographical space: beyond the field-object debate. In: Savanah GA (eds) GIScience, AAG, 52–54 Oct 2000

Chen KS, Yen SK, Tsay DW (1997) Neural classification of SPOT imagery through integration of intensity and fractal information. Int J Remote Sens 18(4):763–783

Curran PJ (1987) Remote sensing methodologies and geography. Int J Remote Sens 8:1255–1275

Durand N, Derivaux S, Forestier G, Wemmert C, Gancarski PO, Boussaid D, Puissant A (2007) Ontology-based object recognition for remote sensing image interpretation. In: IEEE international conference on tools with artificial intelligence, Patras, Greece, pp 472–479

Gahegan M (2001) Visual exploration in geography: analysis with light. In: Miller HJ, Han J (eds) Geographic data mining and knowledge discovery. Taylor & Francis, London, pp 260–287

Gao J, Skillcorn D (1998) Capability of SPOT XS data in producing detailed land cover maps at the urban–rural periphery. Int J Remote Sens 19(15):2877–2891

Gomez B, Jones JP III (eds) (2010) Research methods in geography: a critical introduction. Wiley-Blackwell, West Sussex

Gower ST, Kucharik CJ, Norman JM (1999) Direct and indirect estimation of leaf area index, f(APAR), and net primary production of terrestrial ecosystems. Remote Sens Environ 70(1):29–51

Guba EG, Lincoln YS (1994) Competing paradigms in qualitative research. In: Denzin NK, Lincoln YS (eds) Handbook of qualitative research. Sage, CA

Henning E, Van Rensburg W, Smit B (2004) Theoretical frameworks. In: Henning E, Van Rensburg W, Smit B (eds) Finding your way in qualitative research. Van Schaik Publishers, Pretoria

Holloway I (1997) Basic concepts for qualitative research. Blackwell Science, Oxford

Irny SI, Rose AA (2005) Designing a strategic information systems planning methodology for Malaysian institutes of higher learning (isp-ipta). Issues Inf Syst 6(1):325–331

Jensen JR (2005) Introductory digital image processing: a remote sensing perspective. Prentice-Hall, Upper Saddle River

Jensen JR (2006) Remote sensing of the environment: an earth resource perspective, 2nd edn. Prentice Hall, Upper Saddle River

Johnson LF (2001) Nitrogen influence on fresh-leaf NIR spectra. Remote Sens Environ 78(3):314–320

King RB (1994) The value of ground resolution, spectral range and stereoscopy of satellite imagery for land system and land-use mapping of the humid tropics. Int J Remote Sens 15(3):521–530

Kosso P (2011) A summary of scientific method. Springer, Heidelberg

Kuhn TS (1996) The structure of scientific revolutions, 3rd edn. University of Chicago Press, Chicago and London

Langford M, Bell W (1997) Land cover mapping in a tropical hillsides environment: a case study in the Cauca region of Colombia. Int J Remote Sens 18(6):1289–1306

MacDonald J (2002) The earth observation business and the forces that impact it. Earth Obs Bus Network

Miller HJ, Han J (2001) Geographic data mining and knowledge discovery. Taylor & Francis, London

Patton MQ (1990) Qualitative evaluation and research methods, 2nd edn. Sage, Newbury Park

Peddle DR, Hall FG, Ledrew EF (1999) Spectral mixture analysis and geometric-optical reflectance modeling of boreal forest biophysical structure. Remote Sens Environ 67(3):288–297

Prenzel B (2004) Remote sensing-based quantification of land-cover and land-use change for planning. Prog Plann 61:281–299

Quenzel H (1983) Principles of remote sensing techniques. In: Camagni P, Sandroni S (eds) Optical remote sensing of air pollution. Elsevier Science, Amsterdam, pp 27–43

Ray TW, Murray BC (1996) Non-linear spectral mixing in desert vegetation. Remote Sens Environ 55(1):59–64

Salem BB, El-Cibahy A, El-Raey M (1995) Detection of land cover classes in agro-ecosystems of northern Egypt by remote sensing. Int J Remote Sens 16(14):2581–2594

Schott J (1997) Remote sensing: the image chain approach. Oxford University Press, New York

Shaw IGR, Dixon DP, Jones JP III (2010) Theorizing our world. In: Gomez B, Jones JP III (eds) Research methods in geography: a critical introduction. Blackwell Publishing, West Sussex, pp 9–25

Shuttleworth M (2008) Drawing conclusions. Experiment resources, Online. URL: http://www.experiment-resources.com/drawing-conclusions.html

Silva MPS, Câmara G (2004) Remote sensing image mining using ontologies, Online. URL: http://www.dpi.inpe.br/~mpss/artigos/ImageMining2004.pdf

Smith B (1999) Ontology: philosophical and computational. Unpublished manuscript. URL: http://wings.buffalo.edu/philosophy/faculty/smith/articles/ontologies.htm

Verbyla DL, Richardson CA (1996) Remote sensing clearcut areas within a forested watershed: comparing SPOT HRV panchromatic, SPOT HRV multispectral, and landsat thematic mapper data. J Soil Water Conserv 51(5):423–427

Zarco-Tejada P, Miller J (1999) Land cover mapping at BOREAS using red edge spectral parameters from CASI imagery. J Geophys Res 104(D22):27921–27933

Zarco-Tejada PJ (2000) Hyperspectral remote sensing of closed forest canopies: estimation of chlorophyll fluorescence and pigment content department of physics. York University, Toronto

Zarco-Tejada PJ, Miller JR, Noland TL, Mohammed GH, Sampson PH (2001) Scaling-up and model inversion methods with narrowband optical indices for chlorophyll content estimation in closed forest canopies with hyperspectral data. IEEE Trans Geosci Remote Sens 39(7):1491–1507

Zhang J, Hsu W, Lee M (2002) Image mining: trends and developments. Kluwer Academic, Dordrecht

Chapter 3
Collection of Data

Abstract It was stated earlier that research method involves collection and analysis of data. Once the problem has been stated and the theories have been formed in the research methodology, it is necessary to collect the data, both in situ and remotely sensed, in order to progress toward a solution. If data is to be useful, it must be collected properly. Whatever the logic or research-type used, every problem will have different data requirements. This chapter is aimed to discuss the data and their collection/selection methods and related issues. First it will discuss the factors influencing the selection of remote sensing data for different types of applications; and then the ground truth and other ancillary data will also be addressed. However, the discussion in this chapter will not focus on instrumentations/sensors or scanning/imaging techniques to capture the remote sensing imagery. One may refer any standard remote sensing textbook for these topics.

Keywords Remote sensing • Research • Data • Selection • In situ • Ground truth • Ancillary • Resolution • Optical • Thermal • Microwave • Radar • Passive • Active

3.1 Data for Remote Sensing Research

A researcher should know what sort of data is needed before setting out to collect it. While there may be situations which dictate either in situ or remotely sensed data, many situations will require the researcher to collect both types of data (Jensen 2006). Other than the in situ data, remote sensing research requires many types of ancillary data, e.g., census data, topographical maps, and data from existing geographic information system (GIS).

Required data in any scientific research can be classified into two classes—primary and secondary. Primary data are collected by the investigator conducting the research. Secondary data are collected by someone other than the researcher. Important to realize that remote sensing data are generally not collected by the researcher her/him-self or by their team. Rather, the researcher selects the appropriate data from the available sensors. These data are collected by the respective agencies responsible to maintain and operate the remote sensing sensors. However, collection of ground truth is

B. Bhatta, *Research Methods in Remote Sensing*, SpringerBriefs in Earth Sciences, 43
DOI: 10.1007/978-94-007-6594-8_3, © The Author(s) 2013

primarily to be conducted by the researcher. Other ancillary data are of secondary in nature—collected by people/organizations other than the researcher or their team. However, in remote sensing research, primary data are the remote sensing imageries, although they are not collected by the researcher. Remote sensing data are primary because they form the basis of analysis in such researches. In situ and other ancillary data are categorized as secondary, because they supplement the research. Common sources of secondary data for remote sensing research include ground truth, topographical and other conventional maps, GIS data layers, digital elevation models, censuses, surveys, organizational records, and data collected through other quantitative or qualitative research.

3.1.1 In Situ Data

Remotely sensed data is being used in numerous fields and for a wide variety of applications. Consequently, the collection of in situ data may take the form of field sampling, laboratory sampling, or some combination of both. In situ data are widely used to provide a ground truth for the calibration of remote sensing imageries and/or validation of retrievals from the remote sensing imageries. In remote sensing, ground truth is just a jargon term for at-surface or near-surface observations. The term in its simplest meaning refers to 'what is actually on the ground that needs to be correlated with the corresponding features in the remote sensing image'. Scientific claims must be based on observable evidence; this evidence is ground truth in many instances. Ground truth is especially important in order to relate image data to real features and materials on the ground. The collection of ground-truth data also enables calibration (e.g., atmospheric correction, geometric correction, sensor design, etc.) of remote-sensing data, and aids in the interpretation and analysis of what is being sensed.

The techniques for these types of data collection should ideally be learned from the physical and natural science courses most related to the specific field of study such as chemistry, biology, forestry, soil science, hydrology, or meteorology. It also involves collection of geographic coordinates of the ground resolution cell and comparing those with the coordinates of the pixel being studied to understand and analyze the location errors and how it may affect a particular study. Due to ease of use and increasing affordability, global positioning system (GPS) receivers are the ideal tool to be used to gather such positional data when needed. Using a GPS receiver, an x, y, and z coordinate (longitude, latitude, and altitude) can quickly be obtained to identify and locate individual samples in relation to remotely sensed data.

3.1.2 Remotely Sensed Data

Important to realize, the format and quality of the remote sensing data varies widely. These variations are dependent upon the resolutions of the sensor (radiometric, spatial, spectral and temporal) (Jensen 2006; Bhatta 2011). Sensors are also unique with regard to what portions of the electromagnetic spectrum they see. Different remote sensing instruments record different segments, or bands, of the electromagnetic spectrum. Therefore, determining the correct spatial, spectral, radiometric, and temporal resolution is crucial in any research. Further, whether the research requires optical, or thermal, or microwave data is similarly crucial to determine.

In many instances, remote sensing research requires integration of multiple data; for example, integration of microwave and optical imageries. Integration of remote sensing imageries will be explained in Sect. 4.2.

3.1.3 Other Ancillary Data

In situ data are one type of ancillary data in remote sensing data analysis. There remain many other types of data that are necessary to be used in conjunction with remote sensing data. Ancillary data is accessory or related to the main topic and not of remote sensing origin. These are data from sources other than remote sensing, used to assist in analysis or to populate metadata. Ancillary data are often very diverse, may include maps, vegetation phenologies, and many kinds of information about human activities in general. Sources of ancillary data may include (not limited to): existing maps (analog/digital), reports and publications, field survey or GPS, and existing GIS data layers.

Types of ancillary data and their impacts on remote sensing data analysis is rather vast and beyond the scope of this book. Researcher of a specific field may refer existing papers relating to that field of study to gain the idea on how these data have been used and benefited.

3.2 Factors Influencing the Selection of Remote Sensing Data

As it has already been mentioned, there are several factors involved in selection of remote sensing data. Among them, the resolution and region of electromagnetic spectrum are of high importance. Following sections address these issues in detail. Other factors will be addressed in the context of these two primary factors.

3.2.1 Resolution

Resolution (or resolving power) is defined as a measure of the ability of a remote sensing system or sensor to distinguish between signals that are spatially near or spectrally similar. Data collection system has four major resolutions associated with it. The major characteristics of an imaging remote sensing instrument are described in terms of its *spatial*, *spectral*, *radiometric*, and *temporal* resolutions. Jensen (2006) or Bhatta (2011) may be referred for a detailed discussion on these resolutions.

3.2.1.1 Spatial Resolution

The spatial resolution or the ground resolution cell size of one pixel as the finite image element is the most important characterisation for a remote sensing image (Bhuyan et al. 2007). The detail discernible in an image is dependent on the spatial resolution of the sensor and refers to the size of the smallest possible feature that can be detected. Spatial resolution of passive sensors depends primarily on their *instantaneous field of view* (IFOV). The IFOV is the angular cone of visibility of the sensor which determines the area on the earth's surface that is 'seen' from a given altitude at one particular moment in time. The IFOV may also be defined as the area on the ground that is viewed by a single instrument from a given altitude at any given instant of time (Jensen 2006; Bhatta 2011).

The information within an IFOV is presented by a picture element in the image plane usually referred to as pixel. For a homogeneous feature to be detected, its size generally has to be equal to or larger than the resolution cell. If the feature is smaller than this, it may not be detectable as the average brightness of all features within that resolution cell will be recorded. However, smaller features may sometimes be detectable if their reflectance dominates within a particular resolution cell allowing sub-pixel detection (Yue et al. 2006; Xian and Crane 2005; Brown et al. 2000; Phinn et al. 2002).

In the early days of remote sensing, aerial photography was conducted to obtain high resolution imageries; because satellite remote sensing could not provide very detailed data. Aerial photography has very long archived data records, while satellite remote sensing for earth observation started in 1972 with the launch of first Landsat satellite. Since 1972, numerous technological improvements have led to the second generation of earth observation satellites, such as advanced Landsat satellites (TM and ETM+), SPOT, and Indian Remote Sensing (IRS) LISS sensors. From 1999 one can distinguish a third generation of Earth observation satellites with very high spatial resolution (IKONOS, QuickBird, OrbView, Geoeye, Cartosat, WorldView etc.). This has led to more and more applications using remote sensing data since the requirements regarding the desired level of detail can be fulfilled either by aerial or satellite-based sensor systems. Nowadays, not only these two platforms are complementary, rather satellite sensors are increasingly dominating many application domains.

Section of resolution and thus a sensor is crucial in every remote sensing research. In a low spatial resolution image, larger ground area makes a *mixed pixel* instead of homogeneous pixel. A mixed pixel is a pixel whose digital number (DN) represents the average energy reflected or emitted by several types of surface present within the area that it represents on the ground; sometimes called a *mixel*. However, very high spatial resolution is also not preferred in several instances. Although

Table 3.1 Application scale for various remote sensing images (Neer 1999; Bhatta 2010)

Pixel size in m	Definition	Platform/sensor[a]	Application scale
0.1–0.5	Extremely high res.	Airborne scanner, aerial photos, geoeye-1 (pan), worldview-1 (pan), worldview-2 (ms)	1:500–1:5,000
>0.5–1	Very high res.	IKONOS (pan), quickBird (pan), orbview (pan)	1:5,000–1:10,000
>1–4	High res.	IKONOS (ms), quickBird (ms), orbview (ms), geoeye-1 (ms), IRS (pan)	1:10,000–1:15,000
>4–12	Medium res.	IRS (pan), IRS (LISS-IV ms), SPOT (pan)	1:15,000–1:25,000
>12–50	Low res.	ASTER, IRS (ms), Landsat-TM/ETM+ (pan, ms), SPOT (ms)	1:25,000–1:100,000
>50–250	Very low res.	Landsat MSS	1:100,000–1:500,000
>250	Extremely low res.	NOAA	>1:500,000

[a] pan: panchromatic; ms: multispectral

higher spatial resolution provides better interpretability by a human observer; but a very high resolution leads to a high object diversity which may end up in problems, e.g., when an automated classification algorithm is applied to the data.

Selection of spatial resolution primarily depends on the level of detail required from the remote sensing data (Anderson et al. 1976). Required level of detail further depends on the application area. For example, urban analysis generally requires more detail than a forest analysis; because the urban area is more heterogeneous than the forest. Therefore, the heterogeneity of landscape is a primary factor of choosing the spatial resolution. Another important factor is application scale. For example, the resolution requirement of a research that estimates global built-up area is essentially different for a similar study that concentrates on a city. In the later case, we need to select higher spatial resolution because it requires more detailed data and thus analysis at larger scale. Table 3.1 lists several remote sensing sensors and their application scales.

3.2.1.2 Spectral Resolution

Not only the spatial resolution, but spectral and radiometric resolutions are also vitally important in remote sensing researches. Many earth-surface features have similar spectral characteristics; which means the objects give similar percent reflectance. This makes the object identification and analysis process more complicated.

The spectral resolution refers to the number of spectral bands, individual bandwidths, and the entire range of electromagnetic spectrum covered by the bands. Different classes of features and details in an image can often be distinguished by comparing their responses over distinct wavelength ranges (Jensen 2006; Bhatta 2011). High spectral resolution is achieved by narrow band widths which, collectively, are likely to provide a more accurate spectral signature for discrete objects than broad band-widths. Spectral resolution not only relates with the dimension of the bandwidth but also the number of bands. Individual bands and their widths will determine the degree to which individual targets can be discriminated on a multispectral image. The use of multispectral imagery can lead to a higher degree of discriminating power than any single band on its own. The ideal solution would be a hyperspectral scanner with a large number of bands each with a small bandwidth of 10 nm. But these remote sensing systems are limited and mainly in the experimental stage. Furthermore, these sensors are recent advancement; therefore, for the analysis of past, hyperspectral images are not available. It is important to realize that a lower spectral resolution does not create a mixed pixel; rather it creates *mixed class*, i.e., pixels representing different objects will belong to a single class. Class is a group of pixels relating to a narrow or broad category of objects over the earth surface.

Often, bands in the multispectral image provide redundant information; that means similar type of reflectance in different bands. Therefore, selection of band is very important for a specific project. For example, we require near-infrared (NIR) band for vegetation analysis, because vegetation provides highest reflectance in this band. For the purpose of discriminating different land surface features, firstly we emphasize the band in which highest intra-band variation is present. Intra-band variation refers to the variations in pixel values of different land surface features within a specific band. For example, to identify the water resources within a forested area NIR band is most appropriate; because, in NIR band, water provides lowest and vegetation provides highest reflectance. As a result, there will be a high contrast between the water and vegetation. Once we select the band with highest intra-band variation, the next band should be chosen with highest inter-band variation. Inter-band variation refers to the variations in pixel values between the first and second chosen bands. For example, we consider the NIR band at first for vegetation analysis because it gives highest reflectance in this band; no other objects provide such high reflectance in this band. The second band, for vegetation analysis, we emphasize the red one; because vegetation gives lowest reflectance in this band. As a result, in the first band we have highest intra-band variations among the objects and in the second band we have highest inter-band variations in comparison to the first band for an object to be analyzed.

3.2.1.3 Radiometric Resolution

The radiometric resolution is defined as the sensitivity of a remote sensing detector to differences in signal strength as it records the radiation flux reflected or emitted from the terrain (Jensen 2006; Bhatta 2011). It defines the number of just discriminable signal levels; consequently, it can have a significant impact on our ability to measure the properties of landscape objects. The radiometric resolution of an imaging system describes its ability to discriminate very slight differences in energy. The finer the radiometric resolution of a sensor, the more sensitive it is to detect small differences in reflected or emitted energy. Lower radiometric resolution is also responsible for mixed class instead of mixed pixel.

However, higher radiometric resolution does not always mean a higher quality image, in particular, while we are interpreting the image visually. This is because human beings can differentiate approximately 40–50 individual shades of grey in a black-and-white image. This means that we shall not be able to differentiate between the images having radiometric resolution of 6-bit (64 variations) and 8-bit (256 variations). Tucker (1979) showed that there was only a 2–3 % gain in distinguishing vegetation types using an 8-bit resolution compared to a 6-bit resolution. Slater (1980) illustrates that the signal-to-noise ratio decreases with the increase of radiometric resolution. Signal-to-noise ratio is a measure used to quantify how much a signal has been corrupted by noise. It is defined as the ratio of signal power to the noise power corrupting the signal (power of signal/power of noise). The higher the ratio, the less obtrusive the background noise is. That means lower signal-to-noise ratio causes more background noise. Then what is the utility of having higher radiometric resolution? This feature increases the overall clarity of the image. Further, while processing the image digitally in a computer, the full radiometric range can be utilized in some cases and a better result can be obtained. Unlike human being, computer can recognise all radiometric variations contained within an image.

3.2.1.4 Temporal Resolution

Temporal resolution refers to how frequently the sensor can capture the images for a specific ground area (Jensen 2006; Bhatta 2011). The temporal resolution is an important consideration for a number of monitoring researches, especially when frequent imaging is required (for instance, to monitor the spread of an oil spill or the extent of flooding). Analysis of multiple-date imagery provides information on how the variables change through time. Spectral characteristics of features may change over time and these changes can be detected by collecting and comparing multi-temporal imagery. For instance, during the growing season, most species of vegetation are in a continual state of change and our ability to monitor those subtle changes using remote sensing is dependent on when and how frequently we collect imagery. Each sensor associates an orbital calendar (refer Bhatta 2011). This orbital calendar can provide us the information about when and how frequently we can collect imagery.

Simonett (1983) argues that with some applications, temporal resolution is an important factor. For instance, to monitor crop growth/stress, image intervals of 10 days would be required, but intervals of one year or more would be appropriate to monitor urban growth patterns. The time factor in imaging is important when

- Persistent clouds offer limited clear views of the earth's surface (often in the tropics).
- Short-lived phenomena (floods, oil slicks, etc.) need to be imaged.
- Multi-temporal comparisons are required (e.g., the spread of a forest disease from one year to the next).

- The changing appearance of a feature over time can be used to distinguish it from near-similar features (wheat/maize).

3.2.2 Region of Electromagnetic Spectrum

Remote sensing can be performed in different regions of electromagnetic spectrum. Optical remote sensing is performed within the optical region (0.3–3.0 μm), thermal remote sensing uses the thermal region (3.0–5.0 and 8.0–14.0 μm), and microwave remote sensing is conducted within the microwave region (1 mm–1 m). Optical sensors generally use the sun as a source of energy (an exception is LiDAR). Thermal and passive microwave remote sensing uses emitted energy from the earth's surface. However, active microwave remote sensing throws artificially generated energy to the earth's surface and then the backscattered energy is recorded by the sensor. The use of different wavelengths (and thereby techniques) is mainly because of different application areas.

3.2.2.1 Optical Region

Optical remote sensing makes use of visible, NIR and short-wave infrared sensors to form images of the earth's surface by detecting the solar radiation reflected from targets on the ground. Different materials reflect and absorb differently at different wavelengths. Thus, the targets can be differentiated by their spectral reflectance signatures in the remotely sensed images. Asrar (1989) presented the applications of optical remote sensing. Optical remote sensing data are perhaps most widely used because of their simple nature and easier analysis techniques. Optical remote sensing has emphasized practical applications such as crop inventories, land use classification, and mineral exploration. However, in the last several years, more emphasis has been placed on answering more fundamental scientific questions associated with the global environment, productivity of the oceans, and the composition and distribution of continental rocks (Goetz et al. 1985). Optical remote sensing data are very useful for geologic, vegetation, hydrologic, oceanographic, and urban applications. Optical sensors, unlike the others, can provide very high spatial resolution. Therefore, they are the obvious choice in studies of higher detail requirement.

3.2.2.2 Thermal Region

Thermal remote sensing records emitted energy from the earth's surface. Thermal imageries can be acquired during the day or night (because the radiation is emitted not reflected) and are used for a variety of applications such as military reconnaissance, disaster management (forest fire mapping), heat-loss monitoring, and several others [refer Prakash (2000)]. For the analysis of earth-surface features, for example geologic applications, thermal data may be collected during a 24-hour period. Even multiple data are collected at different times of the day. Temperature extremes and heating/cooling rates can often furnish significant information about the type and condition of an object. For example, the temperature curve of water throughout a day is distinctive for two reasons. First, its range of temperature is quite small compared to that of soil and rocks. Second, it reaches its maximum temperature an hour or two after the other materials. As a result, the terrain temperatures are normally higher than water during the day; however, in the night water is warmer than other materials. Thermal inertia is a measure of the heat transfer rate across a boundary between two materials, e.g., air/soil. Because materials with high thermal inertia possess a strong inertial resistance to temperature fluctuations at a surface boundary, they show less temperature variation per heating/cooling cycle than those with lower thermal inertia. This implies that materials with high thermal inertia have more uniform surface temperatures throughout the day and night than materials of low thermal inertia. For example, water has higher thermal inertia than dry soil; therefore, temperature fluctuation is less in water. Therefore, while choosing the thermal data, one should consider the thermal properties of the objects under investigation and the time of day when the data should be collected. Another issue is important to remember that spatial resolution of thermal sensors is generally coarser. The amount of energy decreases as the wavelength increases (refer Jensen 2006). Therefore, thermal sensors generally posses larger instantaneous field of view (IFOV) compared to optical sensors, to ensure that enough energy reaches the detector in order to make a reliable measurement. IFOV is a measure of the area viewed by a single detector on a scanning system at a given moment in time. Thus, the spatial resolution of thermal sensors is usually fairly coarse, relative to the spatial resolution possible in the visible and reflective infrared.

Most thermal remote sensing applications, such as geologic and soil mapping, are qualitative in the sense. In such cases it is not usually necessary to know the actual temperature or emissivity of the ground-surface material, but simply to study relative differences in the radiant temperature within a scene. In such cases, thermal infrared images are used to (1) determine the type of material in certain instances based on its thermal emission characteristics, (2) discriminate different land surface features, and/or (3) evaluate if significant changes have taken place in the thermal characteristics of these phenomena through time.

However, some thermal remote sensing applications require quantitative data in order to determine absolute kinetic temperatures, for example, sea-surface temperature mapping. The digital data recorded by a thermal sensor, can be processed and calibrated to produce absolute temperature of the ground-surface material. Hence, calibration means developing correlation to relate sensor output values to the actual temperature of the ground objects. This calibration relationship can be applied to each point (pixel) in the digital image, producing a matrix of absolute temperature value. The output of a thermal sensor is the measured quantity of radiant temperature. It is important to realize that the actual radiant temperature cannot be measured by a remote sensing sensor. Atmosphere modifies the energy and thus the recorded amount varies from the actual radiant temperature of an object. Therefore, we need more sophisticated calibration. The precise form of a calibration relationship varies with the temperature range in consideration and the sensor in use. One may refer Malaret et al. (1985); Artis and Carnahan (1982); Zhang et al. (2006) for further details.

3.2.2.3 Microwave Region

Microwave remote sensing can be conducted as both passive and active. Passive microwave remote sensing is similar in concept to thermal remote sensing. All objects emit microwave energy of very little magnitude. A passive microwave imager detects the naturally emitted microwave energy within its field of view. This emitted energy is related to temperature and moisture properties of the emitting object or surface. Passive microwave radiometers generally record energy in the region between 0.15 and 30 cm (between 1 and 200 GHz), well beyond the thermal infrared region. Because the wavelengths are so long, the energy available is quite small compared to optical wavelengths. Thus, the IFOV must be large to detect enough energy to record the signal. Most passive microwave sensors are therefore characterized by low spatial resolution. The amount of radiation measured at different frequencies and polarizations can be analyzed to produce environmental parameters such as soil moisture content, precipitation, sea-surface wind speed, sea-surface temperature, snow cover and water content, sea ice cover, atmospheric water content, and cloud water content. Unlike optical imagers, microwave imagers can operate day or night through most types of weather.

Active microwave remote sensors create their own electromagnetic energy that is transmitted from the sensor towards the terrain (and is largely unaffected by the atmosphere), interacts with the terrain producing a backscatter of energy, and is recorded by the remote sensor's receiver. The strength of the backscattered signal is measured to discriminate between different targets, and the time delay between the transmitted and reflected signals determines the distance (or range) to the target. Active microwave sensors are generally divided into two distinct categories: imaging and non-imaging. Non-imaging microwave sensors include altimeters and scatterometers. In most cases, these are profiling devices which take measurements in one linear dimension, as opposed to the two-dimensional representation of imaging sensors. Radar altimeters transmit short microwave pulses and measure the round trip time delay to targets to determine their distance from the sensor. Generally, altimeters 'look' straight down at nadir below the platform and thus measure height or elevation. Radar altimetry is used on aircraft for altitude determination, on aircraft and satellites for topographic mapping, for the estimation of the height of the sea surface, etc. Scatterometers are also generally non-imaging sensors and are used to make precise quantitative measurements of the amount of energy backscattered from targets. The amount of energy backscattered is dependent on the surface properties (roughness) and the angle at which the microwave energy strikes the target. Scatterometry measurements over ocean surfaces can be used to estimate the wind speeds based on the sea surface roughness. Ground-based scatterometers are used extensively to accurately measure the backscatter from various targets in order to characterize different materials and surface types.

Imaging sensors can create two-dimensional images by measuring the intensity of backscattered energy. The most common form of imaging active microwave sensors is imaging RADAR. The RADAR is an acronym for RAdio Detection And Ranging, which essentially characterizes the function and operation of a radar sensor. As with passive microwave sensing, a major advantage of radar is the capability of the radiation to penetrate through cloud cover and most weather conditions. Because radar is an active sensor, it can also be used to image the surface at any time, day or night. The two primary advantages of radar are—all-weather see-through capability and day or night imaging. It is also important to understand that, because of the basic difference in the operation of an active radar compared to that of passive sensors, a radar image is quite different than the images acquired in the visible and infrared portions of the spectrum. Because of these differences, radar and optical data can be complementary to one another as they offer different perspectives of the earth's surface providing different information content.

While selecting the radar images, two considerations are of most importance—wavelength (or frequency) and polarization. The appearance of the image varies from band to band. For a given surface, longer wavelengths are able to penetrate more than shorter wavelengths. For example, short wavelength radars (3 cm) are reflected from the top of trees. Long wavelength radar (24 cm) can penetrate the vegetation canopy and travels down to the ground and is reflected back. Intermediate wavelengths (6 cm), in some instances, experience multiple scattering events between the canopy, the branches, and the ground. Therefore, it is possible to discern information about the canopy structure and estimate above-ground biomass by acquiring radar imagery of multiple wavelengths of a forested area. The functional capability of the longer L-band and P-band radar wavelengths is that they can penetrate the earth's surface. Space-borne Imaging Radar-C/X-Band Synthetic Aperture Radar (SIR-C/X-SAR) was a joint US-German-Italian project that used a highly sophisticated imaging radar onboard space shuttle to capture images of the earth's surface. The SIR-C/X-SAR obtained data of the Sahara desert that showed braided channels beneath the surface, and this completely surprised the scientists and engineers involved because although it was theoretically possible, they did not anticipate being able to see more than several metres into the surface.

Other than the penetration property, different wavelength is responsible for bright or dark appearance of an object on the imagery. Rough surface scatters more energy towards the sensor and appears brighter compared to smoother surfaces. Whether a surface appears rough (bright) in a radar image depends to a large extent on the radar wavelength. A particular surface may be rough for a short-wavelength radar but smooth for longer wavelengths. Thus, different types of surface can be distinguished by choosing appropriate wavelengths for imaging. In simple words, a surface appears 'smooth' if the height variations are much smaller than the radar wavelength. When the surface height variations begin to approach the size of the wavelength, the surface appears 'rough'.

Other than the wavelength, polarization of radiation is an essential consideration while selecting radar imagery. Polarization refers to the orientation of the electric field of electromagnetic energy. The polarization of the signal has an effect on the nature and magnitude of the backscatter. In a polarized radar, the antenna can transmit and receive signals in either horizontal (H) or vertical (V) mode. Similarly, the antenna receives either the horizontally or vertically polarized backscattered energy, and some radars can receive both. Once the transmitted energy from radar antenna hits the earth's surface, the polarization is modified. Some amount of the energy changes their polarization (called depolarization) while others do not. The amount of depolarization depends on the interaction at the target. Scattering at the earth's surface does not cause significant change in polarization; therefore, the cross-polarized receiving antenna (H send V receive or V send H receive), in general, receives little energy. The strength of like-polarized returns (H send H receive or V send V receive) from targets is stronger than that of cross-polarized. Therefore, radar imagery collected using different polarization and wavelength combinations can provide different and complementary information about the targets on the surface.

3.3 Factors Influencing the Selection of Ancillary Data

It would be ideal if the use of remote sensing techniques could remove the need to acquire any other type of data. However, ancillary data are generally always required for a full and accurate characterization of ground features and to analyze them from remotely sensed data. As stated earlier, ancillary data may vary at a wide range, however, among them ground truth or in situ data are most important.

3.3.1 Ground Truth Data

The remotely sensed image of radiance or reflectance is usually of little value on its own. In most of the cases, deriving the correct information from the analysis of remotely sensed data requires some ground verification data (Steven 1987). Primarily due to environmental and atmospheric variation through time and space, it is usually necessary to measure the property at several places and obtain a site-specific statistical model of the relation between the remote and direct (ground) data (Dozier and Strahler 1983). The model so obtained can be applied to the remainder of the image pixels to estimate the property of interest over the area covered by the image (Curran and Hay 1986; Dungan et al. 1994). Spatial and spectral in situ data are required, for example, to georeference the imagery and to identify the spectral signatures of specific features.

Spatial ground truth data are normally collected using GPS receivers. For proper registration of the imagery, accurate GPS locations, to within the IFOV or pixel size of the imagery, should be collected. That means, if we are registering an image of 5 m resolution, the accuracy of the GPS receiver should be within 5 m. Actually, in practical sense, the accuracy requirement of a GPS receiver is less than the half of the pixel size. If we assume that a coordinate represents the central location of a pixel, then the error in measuring a coordinate can be accepted within the half of the pixel size. If the error exceeds this measure, the coordinate will fall on another pixel.

Spectral ground truth data are collected with spectral radiometers. Since these readings are being collected close to the object, the radiant flux, or energy contributed to airborne and satellite imagery by atmospheric scattering of light or from pixels adjacent to the pixel of interest is minimized or eliminated (refer Jensen 2006). Collected spectral readings enable the remote sensing analyst to radiometrically correct the remote sensing imagery. Laboratory tested spectral reflectance curves are compared to that of remote sensing data to identify different land-surface features on the remote sensing imagery.

One may refer McCoy (2004) to know how to collect ground reference information in support of remotely sensed data. Generally, ground data should be collected at the same time as data acquisition by the remote sensor, or at least within the time that the environmental condition does not change. Current techniques allow interpretation of remote sensing data to determine standard land cover types (vegetation, water, soil, and rock type) with or without significant field verification. However, in order to determine specific conditions of land cover types such as vegetative health, subsurface contamination, or specific water quality criteria, etc., ground-truth data collection is a necessity. Once it has been determined that ground-truth data are required, the following discussion can help in designing the sampling plan (as explained in Bhatta (2011).

Atmospheric conditions: Ground-truth data are used to correct the image for conditions in the atmosphere that intercept incoming solar radiation, thereby affecting the intensity of reflected energy signals. It is preferable to collect this data on the date and time of the collection of the image; however, as an alternative, atmospheric data collected close to the date of collection under similar atmospheric conditions and at approximately the same solar time, can be substituted. The surface water and vegetation spectral data should also be collected at the imaging time, plus or minus a few hours. This time may be extended to 3 h, plus or minus for days close to the summer solstice. The atmospheric conditions of interest are: temperature, humidity, haze or aerosols, wind direction and speed, incident solar radiation, sun elevation and azimuth, dew, precipitation, etc.

Surface water: In places where surface water bodies are of sufficient size and interest, ground data should be collected to represent each type of water condition (representing varying water depths, temperature, relationship to potential source areas, fresh or salt water bodies, turbidity, etc.). Clear water absorbs most of the solar radiation. Reflectivity increases with the presence of solids, either inorganic or organic, and the presence of certain dissolved species can also alter the appearance on the image. Understanding the cause of different reflectivity and absorption in surface water is important to the final interpretation of the image.

Vegetation: Presuming that the analysis of interest is that of vegetative health/ stress, species composition, or plant community mapping, spectral measurements of the various vegetation types in the field will be required. The spectral signature within a pixel of the image consists of an average of the reflectance of all materials within that pixel. For example, for a spatial resolution of 5 m, the spectral response for a stand of vegetation consists of a combination of the spectra of all vegetation types and the soil, ground litter, etc., within the picture element. Ground truthing consists of obtaining spectrometer readings (or samples) for as many of the categories/classes of vegetation types (monoculture and mixed species) as possible within time or access constraints. These readings are obtained above the canopy and should try to be representative of reflectance detected by the airborne or satellite remote sensor. For instance, these categories could include a grassy field, a stand of pine with only forest litter at ground surface, a mix of eucalyptus, live oak, vines, a fresh marsh with mixed vegetation, etc. Each category should be sampled from above the canopy.

Soil, bare ground, and rock: The data collection for soil measurements, as in vegetation spectral sampling, should strive to represent the various types of soil or contamination conditions. Moist soils are darker, and therefore at sites where water content varies, it is a good idea to measure both wet and dry soil conditions. Wet and dry soils, rock outcrop, and other geologic conditions may also warrant collecting location-specific data.

Dark and light calibration targets: Acquiring spectral measurements of uniform bright and dark calibration target areas within a site is recommended. The minimum size of these target areas should exceed the pixel resolution of the imaging instrument. These target areas (known as test site) should be on flat ground, relatively homogeneous, and as close as possible to white or black, respectively, for a given site. Examples of the types of possible targets that can be used are asphalt, cement, sand, a flat aluminium roof, or bare dirt areas. It is best to acquire spectral measurements of these targets as close to the flight time (or conditions at the time of the flight) as possible. The reflectance of these dark and light targets, and their known spectral response, are used to calibrate the imagery and to correct for atmospheric influences. As the collection of ground data is time-consuming and expensive, it is best to establish a test site for sensor design, calibration and validation, and data correction. The test area should be carefully selected with respect to ease of survey, variety of features present, weather condition, and so on. Sensor errors are removed using the ground-truth data of the test site. A recent ground-truthing effort included erecting portable weather stations. The weather data record identifies any changes in conditions during the field measurement period. In addition, solar radiance can be measured directly during the over-flight. A further consideration is that of sample stratification; that is, whether to collect several readings in a single location (narrow-deep sampling) or to collect few samples at many locations (broad-shallow sampling). Site conditions or project goals may drive this decision. Finally, if the understanding of surface or subsurface contaminant conditions is desired, spectral data should be collected in areas of known contamination as well as locations considered to be background or uncontaminated. Transecting across a known plume from contaminated to background might be a desirable approach.

There are several factors of weather and site that affect spectral measurements in ground truthing; therefore, it is best to minimize their contribution to spectral measurements as much as possible. Following are necessary to be remembered for ground spectral measurements.

Sun angles: The major factor that affects the reflectance is the angle of the sun. It has a large impact on the overall brightness and dominates the minor types of spectral change that we try to measure. Sun angle varies with the time of day and day of year. This is the reason for acquiring field spectra (and image data) near solar noon. Sometimes, it is necessary to remeasure twice to determine the magnitude of change, thereby placing an error estimate around the measurements. This problem makes the rigorous comparison of seasonal data difficult. It is easy to identify some major land cover change (e.g., someone cutting down the trees) without considering the sun angle factor. But it is hard to determine vegetation stress (health variation) between time-1 and 2 types of data without this consideration.

Cloud condition: If there is a partial cloud cover, it creates highly dynamic and variable light intensities and can change the spectral quality of light. Hence, it is almost impossible to collect good field spectral data under cloudy conditions. Often people 'see' the cumulus type clouds but not the wispier and thin cirrus clouds, which actually cause more difficulty in interpretations. Some places almost always have cloud cover, but are important to measure. Calibration of instrument before and after each measurement may reduce this problem. This increases time requirements but improves our data quality for the analysis. Sometimes, compromise on the time of day is required to collect field data. It is better to make measurements early in the day (if afternoon clouds are the issue) and assume a correction for the solar angle.

Aerosol, haze, and water vapour: Aerosols, smoke, and water vapour in the atmosphere also deteriorate the quality of image data. Often, it is not possible to change the dates for the work, so we are stuck with these conditions. There is no efficient way to eliminate the effects of aerosols and smoke. Water vapour can be modelled relatively better in the spectra. Having access to weather station data, micrometeorological data are useful later when analyzing the image data. If a more rigorous calibration of these atmospheric properties is required, it can be measured using a 'sun photometer', sometimes called a 'Regan radiometer'. This is used to measure transmission of direct beam light at different wavelengths through the atmosphere. The measurements are made at multiple sun angles and the data is entered into a calibration program that provides information on optical depth, visibility, ozone, water vapour, etc. Typically, this is done on the day of the over-flight to provide corrections for calibrating surface reflectance in the image data.

Topography: The sun angle and topography collectively determine the amount of light from the sun falls on a slope and the reflected light from another slope as well. The light reflected from another slope can also illuminate a slope, which is actually some extra amount of light in addition to sunlight. It is important to keep this in mind when establishing the sampling design. Some sites only have shade-free access at specific times.

Shadows: To minimize shadows in the field data (shadows cause variability in the overall brightness of the reflected spectra), the best method is to lay out a transect that is in the 'principal plane of the sun'. This implies that it is oriented along a track that is parallel with the direct beam of the sun. If we stand with our back to the sun and orient ourselves in the line of least shadows, we will be in this trajectory. We can record our direction of the transect with a compass or with GPS.

3.3.2 Other Ancillary Data

Ancillary data are of a diverse nature, but have been mainly composed by (other than ground truth data) digital elevation model derived data (slope, aspect), cartographic and topographic data (roads, parcel limits, hydrological networks, etc.) or historical data of land uses/land covers. Information regarding temperature, pluviometry, geology, natural catastrophes, etc., have also been included depending on the case to be analyzed. Topographic information, for example, improves the image classification accuracy when the study is performed on a local scale. On the other hand, climatologic data are more useful on regional or continental scales (Skidmore 1989). Census reports (populations, households) can improve the urban growth analysis. Digital elevation data can be used for topographic and sun angle correction. Many other examples can also be given. However, as stated earlier, types of ancillary data and their impacts on remote sensing data analysis is vast. Selection or collection of these data is not within the scope of this book.

References

Anderson JR, Hardy EE, Roach JT, Witmer RE (1976) A land use and land cover classification system for use with remote sensor data. US Geological Survey, Washington, DC

Artis DA, Carnahan WH (1982) Survey of emissivity variability in thermography of urban areas. Remote Sens Environ 12:313–329

Asrar G (ed) (1989) Theory and applications of optical remote sensing. Wiley, Toronto

Bhatta B (2010) Analysis of urban growth and sprawl from remote sensing data. Springer, Heidelberg

Bhatta B (2011) Remote sensing and GIS, 2nd edn. Oxford University Press, New Delhi

Bhuyan MR, Rajak DR, Oza M (2007) Quantification of improvements by using AWiFS over WiFS data. J Indian Soc Remote Sens 35(1):43–52

Brown M, Lewis HG, Gunn SR (2000) Linear spectral mixture models and support vector machines for remote sensing. IEEE Trans Geosci Remote Sens 38:2346–2360

Curran PJ, Hay AM (1986) The importance of measurement error for certain procedures in remote sensing at optical wavelengths. Photogram Eng Remote Sens 52:229–241

Dozier J, Strahler AH (1983) Ground investigations in support of remote sensing. In: Colwell RN (ed) Manual of remote sensing American society of photogrammetry, 2nd edn. Falls Church, VA, pp 959–986

Dungan JL, Peterson DL, Curran PJ (1994) Alternative approaches for mapping vegetation quantities using ground and image data. In: Michener W, Brunt J, StaOEord S (eds) Environmental information management and analysis: ecosystem to global scales. Taylor and Francis, London, pp 237–261

Goetz AFH, Wellman JB, Barnes WL (1985) Optical remote sensing of the Earth. Proc IEEE 73(6):950–969

Jensen JR (2006) Remote sensing of the environment: an Earth resource perspective, 2nd edn. Prentice Hall, Upper Saddle River

Malaret E, Bartolucci LA, Lozano DF, Anuta PE, McGillem CD (1985) Landsat-4 and landsat-5 thematic mapper data quality analysis. Photogram Eng Remote Sens 51:1407–1416

McCoy RM (2004) Field methods in remote sensing. The Guilford Press, New York

Neer JT (1999) High resolution imaging from space—a commercial perspective on a changing landscape. Int Arch Photogram Remote Sens 32(7C2):132–143

Phinn S, Stanford M, Scarth P, Murray AT, Shyy PT (2002) Monitoring the composition of urban environments based on the vegetation-impervious surface-soil (VIS) model by subpixel analysis techniques. Int J Remote Sens 23:4131–4153

Prakash A (2000) Thermal remote sensing: concepts, issues and applications. Inter Arch Photogram Remote Sens XXXIII(Part B1. Amsterdam):239–243

Simonett DS (1983) The development and principles of remote sensing. In: Simonett DS (ed) Manual of remote sensing American society of photogrammetry, vol I, 2nd edn. Falls Church, VA, pp 1–35

Skidmore AK (1989) Expert system classifies eucalypts forest types using thematic mapper data and a digital terrain model. Photogram Eng Remote Sens 55(10):1449–1464

Slater PN (1980) Remote sensing: optics and optical systems. Addison-Wesley, Reading

Steven MD (1987) Ground truth an underview. Int J Remote Sens 8:1033–1038

Tucker CJ (1979) Red and photographic infrared linear combinations for monitoring vegetation. Remote Sens Environ 8(2):127–150

Xian G, Crane M (2005) Assessments of urban growth in the Tampa bay watershed using remote sensing data. Remote Sens Environ 97(2):203–215

Yue W, Xu J, Wu J, Xu L (2006) Remote sensing of spatial patterns of Urban renewal using linear spectral mixture analysis: a case of central urban area of Shanghai (1997–2000). Chin Sci Bull 51(8):977–986

Zhang J, Wang Y, Li Y (2006) A C++ program for retrieving land surface temperature from the data of landsat TM/ETM+ band 6. Comput Geosci 32(10):1796–1805

Chapter 4
Analysis of Data

Abstract Analysis of data in research is a process of inspecting, cleaning/correcting, transforming, classifying, and modelling the data with the goal of highlighting useful information, suggesting conclusions, and supporting decision making. It also refers to the process of evaluating data using analytical and logical reasoning to examine each component of the data. Data from various sources are gathered, reviewed, and then analyzed to form some sort of finding or conclusion. Remote sensing data analysis has multiple facets and approaches, encompassing diverse techniques under a variety of classes and domains. Further, being an emerging field, new tools and techniques are always being proposed and appreciated in the remote sensing community. It is beyond the scope of this book to document each of them. Therefore, this chapter will not attempt to describe the tools and techniques; rather it will be emphasized on the general discussion of remote sensing data analysis. This chapter is based on concepts rather than tools and techniques; constraints and freedoms will also be addressed in context.

Keywords Remote sensing • Data • Information • Knowledge • Image • Analysis • Data mining • Visual interpretation • Digital processing • Multi concept

4.1 Data Analysis and Data Mining

Data analysis, in general sense of remote sensing, is the processing and interpretation of raw data to extract useful information. It also refers to the understanding of the relationship between interpreted information and the actual status or phenomenon, and to evaluate the situation. *Data mining* is a particular data analysis technique that focuses on modelling and knowledge discovery. This technique, as a process of searching and discovering valuable information and knowledge in large volumes of data, is an emerging research field; it draws basic principles from concepts in databases, machine learning, statistics, pattern recognition and 'soft' computing. In fact, data mining should have been more appropriately named 'knowledge mining from data'. Data mining techniques, such as classification, regression, association, and correlation, were broadly applied in the field of remote sensing. However, more and more sophisticated techniques are always being introduced by the researchers.

B. Bhatta, *Research Methods in Remote Sensing*, SpringerBriefs in Earth Sciences,
DOI: 10.1007/978-94-007-6594-8_4, © The Author(s) 2013

Generally speaking, remote sensing data analysis works on the principle of inverse problem. While the object or phenomenon of interest (the state) may not be directly measured, there exists some other variable that can be detected and measured (the observation), which may be related to the object of interest. The common analogy given to describe this is trying to determine the type of animal from its footprints. For example, in remote sensing, a vegetation species can be identified from the pixel values in different bands, or by interpreting the tone, colour, and crown shape.

Remotely sensed data are analyzed using various image processing techniques and methods. These include both visual (or analog) processing techniques applied to hard copy data such as photographs or printouts or the image displayed on computer monitor, and the application of digital image processing algorithms to the digital data (Jensen 1996). One of the purposes of applying both visual and digital techniques to remotely sensed data is to enable the analyst see the data in several ways. The goal of image processing is to allow the researcher to examine their data from all possible angles, to place entire images in context with their surroundings, and to allow the relationships of individual scene elements to be discovered.

Manual interpretation and analysis is the traditional method of remote sensing for air photo interpretation. Digital processing and analysis is more recent with the advent of digital recording of data and the development of computers. Both these techniques for the analysis of remote sensing data have their respective advantages and disadvantages. Table 4.1 provides a comparison between manual and computer information extraction. Human and computer methods supplement each other, so that they both may offer better results when combined.

Digital processing and analysis may be carried out to automatically identify targets and extract information completely without manual intervention by a human interpreter. However, rarely is digital processing and analysis carried out as a complete replacement for manual interpretation. Often, it is done to supplement and assist the human analyst. A computer system with an interactive graphic display through which humans and computers can interactively work together is called 'a man–machine interactive system'. In most cases, mix of both methods is usually employed while analyzing imagery. In fact, the ultimate decision of the utility and relevance of the information extracted at the end of the process still must be made by humans. Currently, analysis through computer, by a means of human–computer interaction, is more popular than the earlier approach of human interpretation on analog media.

A specialist in remote sensing or image processing generally has the tools that allow him, at least in theory, to configure applications solving complex problems of image understanding. However, in reality, earth observation data analysis is still performed in a very laborious way at the end of repeated cycles of trial and error. To overcome this, several researches are being conducted to introduce several novel ideas. These are broadly known as knowledge-driven information mining that is based on human-centred concepts, which implements new features and functions allowing improved feature extraction, search on a semantic level, the availability of collected knowledge, interactive knowledge discovery, and new visual user interfaces (Datcu and Seidel 2005).

Table 4.1 Comparison between information extraction by human and computer (Bhatta 2011)

Method	Advantages	Limitations
Human (visual image interpretation)	• Interpreter's knowledge are available • Excellent in spatial information extraction	• Time consuming • Individual difference • At-a-time analysis of multiple bands/images is difficult • Serious biased error may introduce due to analyst's own assumption or expectation
Computer (digital image processing)	• Short processing time • Re-productivity • Extraction of physical quantities • Data commonly transmitted or converted to digital format • Analysis of individual points (pixel) • Analysis of multiple bands/ images in a single platform • Handling of large data volumes • Data easily manipulated by multispectral and statistical analysis • Full radiometric utilization • Correction of system related errors • Accuracy assessments • Controlled image output • Rapid hard copy • Database correlation • Instant new or revised map generation by printer or plotter	• Human knowledge is unavailable • Spatial information extraction is poor • High expertise in computer Science is required • More expensive (initially)

Remote sensing data are collected and analyzed to enhance understanding of the terrestrial surface—in composition, in form, or in function. One approach for accomplishing this is to design the analysis process as an iterated composite of several analyst-directed modules. Madhok and Landgrebe (2002) have shown one such modular design for the analysis of remote sensing data.

4.1.1 Visual Image Processing

Visual image interpretation is defined as the extraction of qualitative and quantitative information about the shape, location, structure, function, quality, condition, relationship of and between objects, etc. by using human knowledge or experience. It is the process of identifying objects or conditions on remotely sensed images and inferring their significance (Avery and Berlin 1985, 1992). However, image interpretation requires conscious, explicit effort not only to learn about the subject matter, geographic setting, and imaging systems in unfamiliar contexts, but also to develop our innate abilities for image analysis (Campbell 1996). This is one of the areas of image processing that humans excel at—extracting information from images by combining multiple elements of image interpretation. This is because we are continually processing images in our everyday life. As we walk down the street we see the cars, other people, take note of the weather, etc. All these images are passed to our brain where all of our experiences and learning are used to extract the most pertinent information. Similarly, we are very adept at applying collateral data and personal knowledge to the task of image processing. However, interpretation of remote sensing imageries is much complicated comparing the everyday visual interpretation of our surroundings.

What makes the interpretation of imagery more difficult than the everyday visual interpretation of our surroundings? There are four ways in which remote sensing differs from our real life. (1) Imagery is usually acquired from overhead. (2) Many sensors record imagery beyond the visible portion of the electromagnetic spectrum. Healthy vegetation appears red, rather than green, in a colour infrared image. (3) Imagery may be acquired at unfamiliar resolutions and scales. Familiar objects on a high resolution image may not be recognizable on a coarser resolution image. (4) We lose our sense of depth while viewing a two-dimensional image, unless we can view it stereoscopically so as to simulate the third dimension of height. Indeed, interpretation benefits greatly in many applications when images are viewed in stereo, as visualization (and therefore, recognition) of targets are enhanced dramatically.

A typical image interpretation process includes image reading, image measurement, and image analysis. *Image reading* is an elemental form of image interpretation. It corresponds to simple identification of objects using such elements as shape, relative size, pattern, tone, texture, colour, shadow, and other associated relationships. *Image measurement* is the extraction of physical quantities, such as length, location, height, density, temperature, and so on, by using reference data or calibration data deductively or inductively. *Image analysis* is the understanding of the relationship between interpreted information and the actual status or phenomenon, and to evaluate the situation.

While interpreting the imagery, there are a number of characteristics that enable the viewer to detect, recognize, or even identify objects from the imagery. These recognition elements are: shape, size, pattern, shadow, tone or colour, texture, association, and site (Avery and Berlin 1992; Campbell 1996; Jensen 2006). The extent to which each of these elements is used depends on not only on the area being studied, but the knowledge the analyst has of the study area. For example, if an analyst has little or no knowledge of an area depicted in an image they may use the shape of objects to distinguish manmade objects from naturally occurring ones. The texture of an object is also very useful in distinguishing objects that may appear the same if judging solely on tone [i.e., water and tree canopy may have the same mean brightness values, but their texture is much different (Schott 1997)]. All these interpretation elements are very powerful image analysis tools when coupled with a general knowledge of the site. These elements, combined with the multi-concept of examining remotely sensed data in multiple bands of the electromagnetic spectrum (multispectral), on multiple dates (multitemporal), at multiple scales (multiscale) and in conjunction with other scientists (multidisciplinary), allow us to make a judgment not only as to what an object is, but its significance. Since the multi-concept in remote sensing data analysis is extremely powerful and deserves a detailed discussion it will be addressed separately in Sect. 4.2.

4.1.2 Digital Image Processing

Digital image processing is the application of algorithms on digital images to perform processing, analysis, and information extraction. As the term implies, digital image processing is not only a step in the remote sensing process, but is itself a process which consists of several steps. It is important to remember that the ultimate goal of this process is to extract information from an image that is not readily apparent or is not available in its original form. The steps taken in processing an image will vary from image to image for multiple reasons, including the format and initial condition of the image, the information of interest (i.e., geologic formations vs. land cover), the composition of scene elements, and others. There are four general steps in processing a digital image—pre-processing, enhancement, transformation, and classification (Bhatta 2011).

Pre-processing functions involve those operations that are normally required prior to the main data analysis and extraction of information. Before digital images can be analyzed, they usually require some degree of pre-processing. This may involve radiometric corrections, which attempt to remove the effects of sensor errors and/or environmental factors (Schott 1993). A common method of determining what errors have been introduced into an image is by modelling the scene at the time of data acquisition using ancillary data collected. Geometric corrections are also very common prior to any image analysis. If any type of area, direction or distance measurements is to be made using an image, it must be rectified geometrically if they are to be accurate.

Image enhancement is solely to improve the appearance of the imagery to assist the visual interpretation and analysis. There are numerous procedures that can be performed to enhance an image. Common enhancements include image reduction, image magnification, transect extraction, contrast adjustments (linear and nonlinear), spatial filtering, etc.

Image transformations are operations similar in concept to those for image enhancement. However, unlike image enhancement operations, which are normally applied only to a single channel (band or layer) of data at a time, image transformations usually involve combined processing of data from multiple spectral bands; for example, indexing/rationing, Fourier transformations (applied on individual band), principle components analysis, image fusion, etc. Jensen (1996) considered these processing as enhancement.

Image classification and analysis operations are used to digitally identify and classify pixels in the data. Classification is usually performed on multi-spectral data sets, and this process assigns each pixel in an image to a particular class or theme based on statistical characteristics of the pixel brightness values. Unlike the visual image interpretation that uses all of the elements of visual interpretation, digital image processing primarily relies almost on the pixel value. There has been some success with expert systems and neural networks which attempt to enable the computer to mimic the ways in which humans interpret images. Expert systems accomplish this through the compilation of a large database of human knowledge which the computer draws upon in its interpretations. Neural networks attempt to 'teach' the computer what decisions to make based upon a training data set. Once it has 'learned' how to classify the training data successfully, it is used to interpret and classify new data sets.

4.1.3 Information Output

Once the remotely sensed data has been processed, it must be placed into a format that can effectively transmit the information it was intended to. This can be done in a variety of ways including a printout of the enhanced image itself, and image map, a thematic map, a spatial database, summary statistics and/or graphs. Because there are a variety of ways in which the output can be displayed, a knowledge not only of remote sensing, but of such fields like geographic information system (GIS), cartography, and spatial statistics are a necessity. With an understanding of these areas and how they interact one with another, it is possible to produce output that give the researcher the information needed without confusion. However, without such knowledge it is more probable that output will be poor and difficult to use properly.

4.2 Multi-Concept in Remote Sensing Data Collection and Analysis

In the early days of remote sensing when the only remote sensing data source was aerial photography, the capability for integration of data from multiple sources was limited. Nowadays, with most data available in digital formats from a wide array of sensors, data integration is a common method used for interpretation and analysis. Data integration fundamentally involves the combining or merging of data from multiple sources in an effort to extract better and/or more information. This may include data that are in multiplatform, multistage, multiscaled, multispectral, multitemporal, multiresolution, multisensor, multiphase, multipolarization, etc. [refer Bhatta (2011)]. Many remote sensing investigations include several of the aforementioned 'multi' categories. In essence, by analyzing diverse datasets together, it is possible to extract better and more accurate information in a synergistic manner than by using a single data source alone.

A word that naturally springs from the 'multi' concepts is merging. Data acquired by different platforms, with different sensors, at different resolutions, and during different times will tend to be incompatible in some respects. Most common is geometric: the pixel representing radiometric data in some spectral interval from some area on the ground or in the atmosphere is probably not of the equivalent size for the different sensors that monitor the target, be it the earth or a planetary surface, or the properties of the air above. In order to combine data sets from different sources, some adjustments or shifts in both geometric/geographic and radiometric values are required. Two pixels may partially overlap; they may vary in size. Their radiometric character may require modifications (e.g., correcting for atmospheric effects or for bidirectional reflectance). Thus, to successfully merge, both geometric and radiometric corrections must be applied. Some form of re-sampling is usually necessary. Distortions must be reduced or removed. Rectification to some planimetric standard (for instance, a suitable map projection) has to be incorporated. Moreover, fitting or stretching one image to properly overlay another is often vital, requiring ground control points or tie points.

Advantages of multisensors lead to merging an image produced by one sensor with that of another. Both may cover the same wavelength range but differ in, for instance, resolution or pixel size. Or they may be quite different types of sensors, e.g., radar and optical scanners. Different sensors often provide complementary information, and when integrated together, can facilitate interpretation and classification of imagery. An excellent example of merging multisensory images is the combination of multispectral optical data with radar imagery. These two diverse spectral representations of the surface can provide complementary information. The optical data provide detailed spectral information useful for discriminating between surface cover types, while the radar imagery highlights the structural detail in the image.

Merging of low-spatial resolution multispectral images with high-spatial resolution panchromatic images is very popular to obtain a high-spatial resolution multispectral image. The merging of panchromatic data of higher spatial resolution with multispectral data of lower spatial resolution can significantly sharpen the spatial detail in an image and enhance the discrimination of features. Thermal imagery also benefits from merging with other kinds of images.

Multispectral analysis is the study of data in different spectral bands. Multispectral data are essential for creating colour composites, image classification, indices/rationing, principal component analysis, image fusion, etc. Each band of information in a multispectral data collected from a sensor contains important and unique data. We know that different wavelengths of incident energy are affected differently by each target—they are absorbed, reflected, or transmitted in different proportions. The appearance of targets can easily change over time, sometimes within seconds. In many applications, using information of several multispectral bands ensures that target identification or information extraction is as accurate as possible. The use of multiple bands of spectral information attempts to exploit different and independent 'views' of the targets so as to make their identification as confident as possible. Hyperspectral data, as an extension of multispectral data, are more useful to discriminate between different earth-surface features.

Information from multiple images taken over a period of time is referred to as multitemporal information. Multitemporal may refer to images taken days, weeks, seasons, or even years apart. Images taken in different seasons are often referred to as multiseasonal data. Monitoring land-cover change or growth in urban areas requires images from different time periods. Change detection is one of the most important applications of multitemporal data. In general, change detection involves the application of multitemporal datasets to quantitatively (or visually) analyze the temporal effects of the phenomenon. Various techniques have been developed to improve change-detection accuracy, including image overlay, image differencing, image rationing, spectral-temporal classification, post-classification comparison, change vector analysis, the masking method, principal component analysis, etc. Bhatta (2010) and Singh (1989) provide an overview of these methods.

Another valuable multitemporal tool is the observation of vegetation phenology (how the vegetation changes throughout the growing season), which requires data at frequent intervals throughout the growing season. Satellites are ideal for monitoring changes in the earth over time. The repeat cycles of those are measured either in days or a couple of weeks or so. This facilitates monitoring of crop growth and regional vegetation progression, as well as drought and stress conditions. Clear-cutting and environmental damage are also effectively monitored. The effects and status of flooding can also be assessed; particularly now, as there are a number of different satellites in operation, so that the likelihood of any area being imaged on a given day has increased. Of course, daily weather changes are the mission of most of the meteorological satellites. Long-period changes (in years) are also suited to the steady presence of the stable satellites in orbit. The mapping of growth of cities, population points, and other land-use categories are some straightforward uses. The drying up of lakes and the changes in coastal areas can be followed. Degradation and loss of wetlands generally are slow processes that still can be detected over time, perhaps early enough to reverse the trend.

Multistage remote sensing is a strategy for landscape characterization that involves gathering and analyzing information at several geographic scales, ranging from generalized levels of detail at the national level through high levels of detail at the local scale. In remote sensing, a multistage sampling approach to image acquisition involves acquiring complete coverage of a study area with low-resolution imagery, and additional higher-resolution imagery from a sample of locations. Sampling is a method of estimating a measurement of numerous items by evaluating only a portion of them. There are a wide range of statistical sampling methods suited to different applications. Multistage sampling is one method commonly chosen for natural resource surveys using remote sensing. The inventory of forest resources and crop surveys are common examples. Multistage sampling is a hierarchical approach in which the statistical population is first subdivided into a number of primary sampling units, which is considered the first stage. Some of these units are then randomly selected as the sample of the first stage, and information is collected for every sample unit chosen. These units are then subdivided into a series of secondary sampling units, some of which are selected as the sample of the second stage. This process can be repeated for the third and further stages. In remote sensing, a multistage approach is often applied to image acquisition. Less expensive lower-resolution image coverage is obtained for an entire study area (the first stage). Then more expensive higher-resolution imagery is obtained for selected areas (the second stage). More detailed imagery or surface observations may then be acquired for a sample of these second-stage sites, resulting in third-stage data, and so on.

Multiscale images require a series of images at different scales, taken at the same time. Although simultaneous acquisition is difficult, it is often possible to acquire images from different sources that were taken at approximately the same time, i.e., within a few days of one another. Multiscale images could include satellite-based images, airborne images taken from different flying heights, or using different camera lenses. In general, for interpreting multiscale images, we use the larger-scale images to interpret smaller scale imagery. Alternatively, smaller-scale imagery may be used for reconnaissance purposes and larger-scale imagery for more detailed analysis within selected sub-areas of the smaller-scale image.

Applications of multisource data integration generally require that the data be geometrically registered, either to each other or to a common geographic coordinate system or map base. This also allows other ancillary (supplementary) data sources to be integrated with the remote sensing data. For example, elevation data in digital form, called digital elevation or digital terrain models (DEMs/DTMs), may be combined with remote sensing data for a variety of purposes. The DEMs/DTMs may be useful in image classification, as effects due to terrain and slope variability can be corrected, potentially increasing the accuracy of the resultant classification. The DEMs/DTMs are also useful for generating 3D perspective views by draping remote sensing imagery over the elevation data, enhancing visualization of the area imaged.

Ground-truth activities are an integral part of the multi concept. Supporting ground observations should be obtained from many relevant, but not necessarily interrelated, sources (multisource). The collection of ground-truth data enables calibration (e.g., atmospheric correction, geometric correction, sensor design, etc.) of remote-sensing data, and aids in the interpretation and analysis of what is being sensed.

Combining data of different types and from different sources, such as those described earlier, is the pinnacle of data integration and analysis. In a digital environment, where all the data sources are geometrically registered to a common geographic base, the potential for information extraction is extremely wide. This is the concept for analysis within a digital GIS database. The integration with GIS allows a synergistic processing of multisource spatial data.

4.3   Level of Detail

The level of detail that can be reliably extracted from remote sensing data is a function of the spatial resolution of the image or the scale. In a low spatial resolution image, larger ground area makes a mixed pixel (*mixel*) instead of homogeneous pixel. This problem increases in cases where within a small transect the heterogeneity of land-cover is very high. However, very high spatial resolution is also not preferred in many applications. Although higher spatial resolution provides better interpretability by a human observer; but a very high resolution leads to a high object diversity which

Table 4.2 Different levels of land-use/land-cover classification (Anderson et al. 1976)

Level 1	Level 2	Level 3
1. Urban or built-up land	1.1 Residential	1.1.1 Single family units
		1.1.2 Multi-family units
		1.1.3 Group quarters
		1.1.4 Residential hotels
		1.1.5 Mobile home parks
		1.1.6 Transit lodgings
		1.1.7 Other
	1.2 Commercial and services	
	1.3 Industrial	
	1.4 Transportation, communications, and utilities	
	1.5 Industrial and commercial complexes	
	1.6 Mixed urban or built-up land	
	1.7 Other urban or built-up land	
2. Agricultural land	2.1 Cropland and pasture	
	2.2 Orchards, groves, vineyards, nurseries, and ornamental horticultural areas	
	2.3 Confined feeding operations	
	2.4 Other agricultural land	
3. Rangeland	3.1 Herbaceous rangeland	
	3.2 Shrub and brush rangeland	
	3.3 Mixed rangeland	
4. Forest land	4.1 Deciduous forest land	
	4.2 Evergreen forest land	
	4.3 Mixed forest land	
5. Water	5.1 Streams and canals	
	5.2 Lakes	
	5.3 Reservoirs	
	5.4 Bays and estuaries	
6. Wetland	6.1 Forested wetland	
	6.2 Non-forested wetland	
7. Barren land	7.1 Dry salt flats	
	7.2 Beaches	
	7.3 Sandy areas other than beaches	
	7.4 Bare exposed rock	
	7.5 Strip mines, quarries, and gravel pits	
	7.6 Transitional areas	
	7.7 Mixed barren land	
8. Tundra (a vast treeless plain)	8.1 Shrub and brush tundra	
	8.2 Herbaceous tundra	
	8.3 Bare ground tundra	
	8.4 Wet tundra	
	8.5 Mixed tundra	
9. Perennial snow or ice	9.1 Perennial snowfields	
	9.2 Glaciers	

may end up in problems when an automated classification algorithm is applied to the data. The highest level of information extraction requires the largest scales (1:1,000–1:5,000), the medium level would focus on medium scales (1:10,000–1:25,000), the lowest level of detail focuses on regions requiring only small scales (1:50,000–1:100,000). Table 3.1 lists several remote sensing sensors and their application scales.

The level of detail is also related with the level of classification and analysis. Anderson et al. (1976) has suggested a four-level classification system for land-use/land-cover classification. Level 1 gives a very broad level classification (greater than 1:500,000 scale). With an increasing scale of image, a higher level classification is possible. Table 4.2 gives level 1 and 2 for land-use/land-cover classification system according to Anderson et al. (1976). Further categorization of level 3 and 4 can be made depending on the scale of the image and details required by the analyst. An example of level 3 classification of level 2 residential category is also shown in Table 4.2. As further advances in technology are made, it may be necessary to modify the classification system.

4.4 Limitations of Remote Sensing Data Analysis

It has already been discussed in Chap. 3 that remote sensing data are challenged by spatial, spectral, radiometric, and temporal resolutions (refer Chap. 3 for detail). Other than these resolution constraints, when dealing with remote sensing multi data, the misregistration problem is a vital issue. Misregistration means registration error between multiple images. In many applications (e.g., image fusion, image subtraction, etc.) we need very accurate image registration that is often very difficult to achieve.

Data generation method is another important issue that includes remote sensing imaging, post-imaging processing (e.g., image classification), and generation of ancillary data. Different sensors generate different type of images that are often to be considered in temporal analysis that result in different images to be compared. Different image processing techniques (or algorithms) may also generate different results for the same image. Different classification schemes (level of classification) also generate different resultant maps. Further, the generation method of ancillary data that are often required for analytical or other purposes (e.g., validation) may also vary in a wide spectrum. Use of these varying data generation methods is a common practice in remote sensing data analysis; because data generation methods and resulting data are not designed to be consistent with different datasets. However, the question arises regarding the validity of directly comparing spatial data derived from different data generation processes. Depending on the data generating process used (for example, the spatial resolution, spatial accuracy, classification schemes and the rules used to define various classes), data may or may not be comparable.

Next issue is classification accuracy. Although classified remote sensing imagery at a range of scales has long been used in many applications, classification of land-covers within a heterogeneous (e.g., urban) landscape is a difficult task. In the heterogeneous areas, generally there are many landscape features present in a small transact. This results in virtually every pixel of even high spatial resolution imagery being a mixture of a vast range of different surfaces (mixed pixel problem). In addition, identifying accurately which pixel matches which area on the ground can represent a difficult problem, leading to registration errors. Error and uncertainty in remote sensing based land-cover maps also represent a major drawback to operational application. Inaccuracies, which may be produced when converting spectral reflectance values to land-cover classes, are further compounded when inferring land-use from such land-cover maps. The classification capabilities of remote sensing data mainly depend on the spectral contrast between the classes of interest and the spectral resolution of the sensor. The lower the spectral separability of classes to be determined means less accuracy in the classified land-cover map.

An overall classification accuracy of 85 % is commonly considered sufficient for a remote sensing data product. Post-classification comparison of two such temporal images will produce an accuracy around 72 % (0.85 × 0.85 × 100) that may not be acceptable in many of the instances. Furthermore, a generalised class definition in the classification process may result in a representation of the landscape that is too homogenous. In contrast, if the landscape classification is too detailed, relevant structures may get lost in a highly heterogeneous pattern. Important to realise, the classification accuracy of the remote sensing data usually decreases as more classes are derived.

Finally, analysis and interpretation of remote sensing data is a difficult task. It requires understanding of theories on how the instrument is making the measurements, what the possible uncertainties may be involved in measurements, and essentially need sufficient knowledge of the phenomena being analysed.

4.5 Converting Remote Sensing Data into Information

Remote sensing provides a new source of information that cannot be easily obtained in other ways and that promises both economic and social benefits. This promise will require a better understanding of cost-effective ways to realize potential useful applications. Utility of remote sensing will come not from the data itself but rather from the information that can be derived from the data. Turning data into useful information is central to technology transfer and the development of successful applications. Till date new applications of remote sensing data have been developed largely by individuals or organizations that already possessed both the necessary technical expertise and an understanding of potential uses of the data. Remote sensing data may initially appear complicated and possibly even irrelevant to potential end users who make policy and management decisions. Such users need easily understandable information that can be used to address economic, social, environmental, and other policy questions. For this reason, research to enable interpretation of the data, and transformation of remote sensing data into usable information, are critical steps in the development of applications.

To enjoy widespread use, remote sensing data must be made accessible to information consumers who may not have the technical expertise currently required to use such data. Past approaches to applications development have stopped short of realizing remote sensing's full potential (refer SCSAC 2001). The demand for applications will be driven by requirements for information rather than by the technical capabilities of the end users. Unlike those who developed the first applications of remote sensing, many new application users are likely to have little, if any, knowledge of remote sensing technology or how it is employed to derive information. They will be concerned instead with the accuracy and timeliness of the data and with its relevance for specific tasks and decisions.

Critical element in producing information of value is processing, which involves two steps: pre-processing and the conversion of data to information. Pre-processing turns raw data into accurately calibrated measures of precisely located physical variables such as reflectance, emittance, temperature, and backscatter. The knowledge base underlying this step is usually well developed, although research may be required for developing specific applications. However, the scientific knowledge base to support the conversion of data to information is far less developed. Transforming technical data into a form that is meaningful to nontechnical users—a process often including either the integration of remote sensing data with other types of data or scientific research to characterize the data (or both)—is highly dependent on the information requirements of applied users and on the skills of technical experts. The diversity of end users' information needs that might be met by the same initial set of physical variables can be depicted in the conceptual diagram given by SCSAC (2001) (Fig. 4.1), which illustrates several simultaneous data-to-information conversion processes.

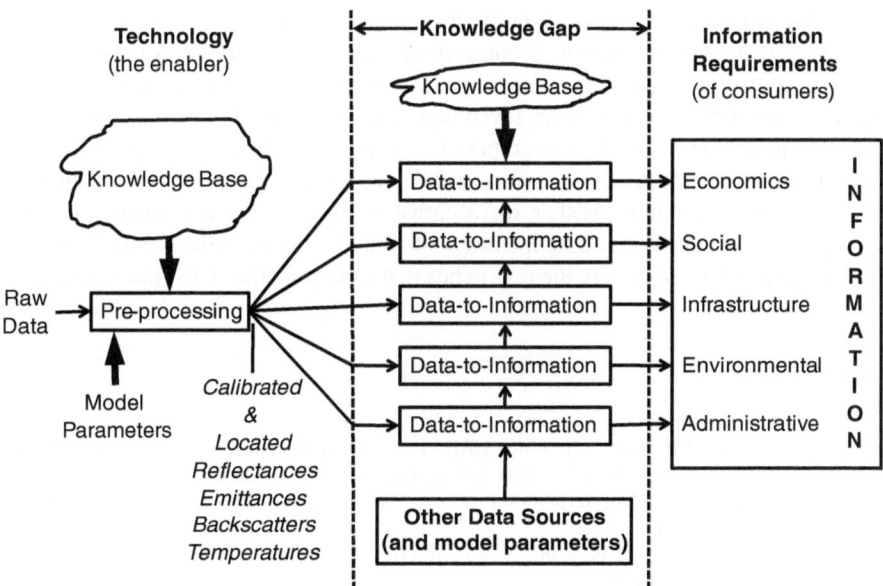

Fig. 4.1 Processing data into information (SCSAC 2001)

Data must be transformed into information and knowledge if the goal of developing successful operational applications of remote sensing data is to be met. On one side of the gap are the scientists, engineers, and technologists who construct and operate instruments to measure parameters in the earth system using spacecraft and aircraft. On the other side are actual and potential end users (the information consumers) who have information requirements but know little about using remote sensing technology to satisfy them and, critically, have little if any financial, personal, or institutional motivation to consider such an approach. The presence within an organization of a highly motivated person who can get an organization to recognize the potential benefits of using remote sensing data and applications can be critical to overcoming gaps in communication.

Nevertheless, those who develop sensors, collect and analyze data, and develop products to address scientific or technical questions often have few opportunities to communicate with prospective users who lack essential technical expertise. This lack of communication constitutes a significant barrier to technology transfer. Bridging the knowledge gap will depend not only on improved communication among technologists and users (refer SCSAC 2001), but also on research focused on converting data into information. A key barrier to transforming data and images into meaningful information is the limited understanding of how to convert measurements made from space into information of ecological, economic, social, infrastructure, environmental, or administrative value. Improving this knowledge base requires the involvement of those who are knowledgeable about the physics of remote sensing and the technologies that support it. In addition, both social and natural scientists could provide and integrate complementary data for use in the transformation of raw remote sensing data into usable information, and researchers could also refine or extend the utility of an existing remote sensing application.

However, the current model of conducting scientific research and publishing the results in peer-reviewed journals that emphasize original new work can discourage the pursuit of research on applications, given that many scientific disciplines and universities often judge the merit of a researcher or a project by publication alone. Moreover, funding organizations generally issue grants for original research rather than work on applications, creating another disincentive to turning remote sensing research into applications. Widespread applications of remote sensing will essentially remain 'remote' unless we emphasize the researches that can dilute the stated knowledge gap.

References

Anderson JR, Hardy EE, Roach JT, Witmer RE (1976) A land use and land cover classification system for use with remote sensor data. US Geological Survey, Washington DC
Avery TE, Berlin GL (1985) Interpretation of aerial photographs, 4th edn. Burgess Publishing Company, Minneapolis
Avery TE, Berlin GL (1992) Fundamentals of remote sensing and airphoto interpretation, 5th edn. Prentice Hall, New Jersey
Bhatta B (2010) Analysis of urban growth and sprawl from remote sensing data. Springer, Heidelberg

Bhatta B (2011) Remote sensing and GIS, 2nd edn. Oxford University Press, New Delhi

Campbell JB (1996) Introduction to remote sensing. The Guilford Press, New York

Datcu M, Seidel K (2005) Human-centred concepts for exploration and understanding of earth observation images. IEEE Trans Geosci Remote Sens 43(3):601–609

Jensen JR (1996) Introductory digital image processing: a remote sensing perspective, 2nd edn. Prentice-Hall, New Jersey

Jensen JR (2006) Remote sensing of the environment: an earth resource perspective, 2nd edn. Prentice Hall, New Jersey

Madhok V, Landgrebe DA (2002) A process model for remote sensing data analysis. IEEE Trans Geosci Remote Sens 40(3):680–686

Schott JR (1993) Methods for estimation of and correction for atmospheric effects on remotely sensed data. In: Presented at SPIE's OE/aerospace and remote sensing 1993 1968, vol 51. Orlando, Florida, pp 448–482

Schott JR (1997) Remote sensing: the image chain approach. Oxford University Press, New York

SCSAC (Steering Committee on Space Applications and Commercialization, National Research Council) (2001) Transforming remote sensing data into information and applications. National Academies Press, Washington

Singh A (1989) Digital change detection techniques using remotely sensed data. Int J Remote Sens 10(6):989–1003

Chapter 5
Research Design

Abstract This chapter deals with the research design and its parts. A research design is the arrangement of conditions for collection and analysis of data. It facilitates the smooth flow of various research processes. It would result in more accurate results with minimum usage of time, effort and money. The research design is considered as an 'outline' or 'conceptual structure' or 'blueprint' for research. Some basic questions that need to be addressed are: which questions to study, which data are relevant, what data to collect, and how to analyze the data. These basic questions may raise some other questions. These questions may be theoretical or practical in nature. The best design depends on the research question as well as the orientation of the researcher. Research design has, in general, the following four parts: (1) sampling design, (2) observational design, (3) analytical design, and (4) operational design. These parts of research design are cross-cutting and one complements others.

Keywords Remote sensing • Research • Design • Philosophy • Sampling • Observation • Theory • Measurement • Variable • Relationship • Validity • Reliability

5.1 Research Design

Research design is the conceptual structure within which research is conducted; it constitutes the blueprint for the collection, measurement and analysis of data. As such the design includes an outline of what the researcher will do from writing the hypothesis and its operational implications, to the final analysis of data. More explicitly, the design decisions address the following questions: (1) What is the study about? (2) Why is the study being made? (3) Where will the study be carried out? (4) What type of data is required? (5) Where can the required data be found? (6) What periods of time will the study include? (7) What will be the sample design? (8) What techniques of data collection will be used? (9) How will the data be analyzed? (10) In what style will the report be prepared?

B. Bhatta, *Research Methods in Remote Sensing*, SpringerBriefs in Earth Sciences,   77
DOI: 10.1007/978-94-007-6594-8_5, © The Author(s) 2013

Keeping in view the aforementioned design decisions; one may split the over-all research design into the following parts: (1) sampling design, (2) observational design, (3) analytical design, and (4) operational design. Sampling design relates to the methods of selecting items to be observed in the research (refer Sect. 5.4). Observational design deals with the conditions under which observations are made (refer Sect. 5.5). Analytical design concerns about how the gathered data are to be analyzed (refer Sect. 5.6). Operational design deals with the techniques by which the procedures specified in the sampling, analytical and observational designs can be carried out (refer Sect. 5.7).

From the preceding discussion, we can list the important features of research design as follows: (1) It is a plan that specifies the sources and types of infor-mation relevant to the research problem. (2) It is a strategy specifying which approach will be used for gathering and analyzing the data. (3) It also includes the time, and cost budgets since most studies are done under these two constraints.

In brief, research design must contain at least the following features: (a) a clear statement of the research problem; (b) procedures and techniques to be used for gathering information; (c) the population to be studied; and (d) methods to be used in processing and analyzing the data.

5.2 Functions of Research Design

Regardless of the type of research design selected by the research, all plans per-form one or more functions outlined and discussed in this section. The number of functions performed by any design largely depends upon its sophistication, cou-pled with the researcher's concerns. First of all, a research design acts as a *blue-print*. Further, it also associates *directional* and *anticipatory* functions.

Perhaps the most important function of research designs is that they provide the researcher with a blueprint for st udying research questions. Without this blueprint or plan, a researcher faces several obstacles in every step of their research. To min-imize the research problems, there are several decisions one should make before beginning the research. For example, if one chooses to study the urban growth of a city, some possible considerations might be: (1) description of the city about which the researcher seeks information (this gives brief idea), (2) defining the geographic extent of the study area, (3) deciding the time span under analysis (temporal aspect), (4) selection of remote sensing data (consideration of spatial, spectral, and temporal aspects), (5) deciding on ground sampling methods, (6) selection of other ancillary data, (7) possible ways of analyzing the data (selection of conceptual, mathematical, statistical, and procedural methods and models), (8) what may be the outcome and its usefulness, etc. These problems are given strong consideration in a research proposal, prospectus, or study outline that many inves-tigators elect to construct in advance of their research.

Similar to the blueprint, directional function is also equally important in designing a research. Research designs dictate boundaries of research activity and enable the investigators to channel their energies in specific directions. Without the delineation of research boundaries and/or objectives, a researcher's activities in a single project could be virtually endless. With clear research objectives in view, researcher can proceed systematically towards the achievement of certain goals. The structure provided by the research plan enables the investigator to reach closure and consider any given research completed.

The third function of a research design is anticipatory. It enables the researcher to anticipate potential problems in the implementation of the study. It is customary for researchers to review current literature central to the topic under investigation. In the course of the literature review, they may learn about new or alternative approaches to their problems. At the same time they can acquire information concerning what can reasonably be expected to occur in their own investigation. Many articles in the professional journals, as well as specialized monographs, include suggestions for further study. More importantly, many authors provide criticisms of their own work so that future investigations of the same or similar topics may be improved. In addition, the design can function to provide some estimate of the cost of the research, possible measurement problems, and the optimal allocation of resources such as assistants (manpower) and material.

5.3 Features of Research Design

Research design is needed because it facilitates the smooth sailing of the various research operations, making research as efficient as possible, yielding maximal information, with minimal expenditure of effort, time, and money. A good research design is often characterized by adjectives like flexible, appropriate, efficient, economical, and so on. Generally, the design which minimizes bias and maximizes the reliability of the data collected and analyzed is considered as good design. The design which gives the smallest experimental error is supposed to be the best design in many investigations. Similarly, a design which yields maximal information and provides an opportunity for considering many different aspects of a problem, is considered most appropriate and efficient design in respect of many research problems. Thus, the question of good design is related to the purpose or objective of the research problem, and with the nature of the problem to be studied. A design may be quite suitable in one case, but may be found wanting in one respect or other, in the context of some other research problem. One single design cannot serve the purpose of all types of research problems.

A research design appropriate for a particular research problem, usually involves the consideration of the following factors: (1) the means of obtaining information, (2) the availability and skills of the researcher and her/his assistants (if any), (3) the objective of the problem to be studied, (4) the nature of the problem to be studied, (5) the availability of time and money for the research work. If the research study happens to be an exploratory or a formulative one, where the major emphasis is on the discovery of ideas and insights, the research design must be flexible enough to permit the consideration of many different aspects of a phenomenon. But, when the purpose of a study is to accurately describe a situation, or an association between variables (or, in what are called descriptive studies), accuracy becomes a major consideration and a research design which minimizes bias and maximizes the reliability of the evidence collected is considered a good design. Studies involving the testing of a hypothesis of a causal relationship between variables require a design which will permit inferences about causality in addition to the minimization of bias and maximization of reliability. But, in practice, the most difficult task is to put a particular study in a particular group, for a given research may have in it elements of two or more of the functions of different studies. It is only on the basis of its primary function that a study can be categorized, either as an exploratory or descriptive, or hypothesis testing study, and, accordingly, the choice of a research design may be made in case of a particular study. Further, the availability of time, money, the skills of the research staff, and the means of obtaining the information must be given due emphasis while working out the relevant details of the research design, such as experimental design, survey design, sample design, and the like.

5.4 Sampling Design

Sampling means measuring a small portion of something and then making a general statement about the whole thing. Sampling is concerned with the selection of a subset of individuals from within a population to estimate characteristics of the whole population. Sampling makes possible the study of a large, heterogeneous population. Researchers rarely survey the entire population, because the study area may be too large or the population to be studied is unlimited. Sampling makes possible this kind of study because in sampling only a small portion of the phenomena may be involved in the study, enabling the researcher to reach all through this small portion. The three main advantages of sampling are that the cost is lower, data collection is faster, and since the data set is smaller it is possible to ensure homogeneity and to improve the accuracy and quality of the data.

5.4.1 Remote Sampling

Remote sensing is a sampling method of data collection. It involves sampling and quantization of reflectance from the sampling ground area. For example, Landsat-TM represents a single pixel value (in a band) for a ground area of 30 × 30 m. It cannot capture every detail from this area. With the advent of new generation sensors the sampling rate has essentially been increased; for example WorldView-2 panchromatic sensor provides information for 0.46 × 0.46 m ground area. However, still it cannot provide every detail of land surface. The sampling interval in electromagnetic spectrum is also coarser in panchromatic or multispectral images. Remote sensing sensors cannot provide reflectance value for every wavelength; for example WorldView-2 panchromatic sensor provides a single value for the wavelength range of 0.450–0.800 μm. Although, this sampling rate has been increased in hyper-spectral sensors, they are also incapable to capture every wavelength. However, it is capable for continuous sampling of narrow intervals of the spectrum, as little as 0.01 μm. One needs to remember that, increase in spectral sampling rate essentially decreases spatial sampling rate, and vice versa. Temporal sampling rate is another issue in remote sensing. Temporal sampling rate refers to the temporal resolution of the sensor. Although the time is a continuous phenomenon, remote sensing cannot provide data for each and every time. However, new scanning techniques have improved the rate of temporal sampling, especially in scene specific cases.

From the previous discussion it is clear that remote sensing is a sampling technique in spatial, spectral, and temporal dimensions. Advancement in technology has potentially increased these sampling rates. However, as mentioned in Chap. 3, (Sect. 3.1), remote sensing data are not generally collected by the researcher her/him-self. These data are collected by the respective agencies responsible to maintain and operate the remote sensing sensors. Therefore, resampling of image data is often required to serve the application/research purpose. The word 'resampling' actually refers to spatial resampling, not spectral or temporal. Spectral and temporal resampling is near to impossible.

Remote sensing imageries contain several flaws or deficiencies in their raw form. To remove/reduce the flaws or deficiencies spatial resampling is performed; for example, in case of georeferencing, or to make the pixels square shaped (from rectangular sampling), or to match the spatial resolution of one image to the other. Interpolation techniques (nearest neighbour, bilinear, cubic convolution) are used to resample the image data. One should be aware of the advantages and disadvantages of these resampling methods [refer Bhatta (2011)].

5.4.2 Ground Sampling

Although remote sensing data are generally not collected by the researcher, ground sampling requires direct involvement of the researcher or their team. Inferring conditions about the earth's surface by using remotely sensed images almost always requires the use of reference, or 'ground truth' data. Ground-based measurements typically involve collecting measurements of the phenomena or target being remotely sensed and can range from employing physical field checks to aerial photography (Lillesand et al. 2004). More commonly, they include physical and chemical measurements with geographic positions (latitude/longitude) and observations for comparison with remotely sensed imagery. Generally, ground truth is used to assist with (1) image analysis and interpretation (e.g., image classification) of remotely sensed imagery, (2) remote sensor calibration, and (3) accuracy assessment of image analysis results (Lillesand et al. 2004).

Much of the emphasis on collecting ground truth is for verification and assessment of image analysis; however, there are no universally accepted standards for assessing accuracy (Congalton and Green 2009). Considerations for assessing accuracy in the collection of ground truth should include such topics as the distribution of the phenomena being mapped, sample/field size, number, type, heterogeneity and distribution of phenomena under investigation, frequency of collection, project budget, and consistency and objectivity in measurement and collection (Congalton and Green 2009). Researchers in the 1970s began to introduce simple techniques for accuracy assessment (Ginevan 1979), followed by more detailed efforts described in Congalton et al. (1983). Furthermore, some guidance and examples for statistically sound approaches in determining sample size are available (Hord and Brooner 1976; van Genderen and Lock 1977; Hay 1979; Rosenfield et al. 1982; Congalton 1988). Also, consideration for choosing the appropriate sampling strategy is described in Ginevan (1979), Fitzpatrick-Lins (1981), Stehman (1992) and Storie and Bugden-Storie (2010).

When a remote sensing research is under process, it is mandatory to have sufficient ground truth data to test not only the accuracy of the final image analysis output, but also the intermediate steps in that process. General guidance for ground truth collection in remote sensing research is extremely limited, as it is dependent on specifics of the technology being developed. Although much work has been done to develop procedures for accuracy assessment, there is a need to better understand and develop the role of ground truth and collection methods for evolving remote sensing technology.

5.5 Observational Design

The comparison of observational information with theoretical ideas constitutes the core of scientific inquiry and represents the process by which scientific knowledge is generated. This era, referred to as the 'quantitative revolution', shifted the emphasis from simple visual observation and qualitative accounts of observational facts toward the production of quantitative observational 'data' (Rhoads and Wilson 2010). The analysis of relationships among quantifiable variables relied heavily on mathematical and statistical methods, which permitted causal relations to be established through inductive inferences made on the basis of the mathematical/statistical results. More theoretically, quantitative data and the use of mathematical/statistical analysis are being viewed as tools for preserving the objectivity of research inquiry.

5.5.1 Theory and Observation

Research in remote sensing can interestingly be viewed as the interplay between theory and observation. Theory consists of 'ideas' and 'proves' about how the earth surface features are structured, and how they interact one another. As it was mentioned in Chap. 1, the word 'observation' has dual meaning in remote sensing: (1) observation of the earth-surface by using remote sensing sensors/cameras, and (2) observation of earth-surface features (their shapes, sizes, spatial relationships, spectral reflectance characteristics, and so on) on the remote sensing images by the researcher. Other than these two, ground-truthing or ground observation is also made by the researcher. However, in the most general sense, observation involves the sensor to observe the earth's surface in spectro-spatio-temporal scales. The primary objective of observation is to obtain information about the structure and dynamics of the earth-surface features so that this information can be compared with theories. The comparison of observational information with theoretical ideas/proves constitutes the core of scientific inquiry and represents the process by which scientific knowledge is generated. Regardless of the differing opinions that have emerged about the interplay between theory and observation, most researchers accept that this interplay is central to any remote sensing inquiry.

Specific conceptions, in general, are embodied in the different philosophical perspectives on scientific inquiry. Perhaps the most well-known of these perspectives is logical positivism (sometimes known as logical empiricism). In many ways, logical positivism formalized the inductive, empirical approach to inquiry that marked the beginnings of modern science in the sixteenth century. It conforms to the classical view of the scientific method in which scientific inquiry begins with observation, and theories develop through attempts to organize facts into systematic explanatory relationships. According to positivism, observations are theoretically neutral. This theoretically neutral information derived from observation can, through the process of hypothesis testing, be used to adjudicate among competing theories.

By contrast, post-positivist perspectives on science emphasize that observation is theory-laden; that is, the information content of all observations is at least partly influenced by a 'priori' theoretical commitments (Rhoads and Thorn 1996). A priori knowledge is independent of experience, whereas 'a posteriori' knowledge is dependent on experience, and these two forms of knowledge differentiate the epistemological notions of 'inductive' and 'deductive' reasoning. Differences between these perspectives depend on the degree to which observation is viewed as 'theory-dependent'. If observation is highly dependent on theory, especially a theory under test, information derived from the observation may be biased and unhelpful in evaluating the theory.

In case of remote sensing, observation is viewed not as information obtained directly through the senses (vision, hearing), as the logical positivists maintained, but as data gathered using elaborate theory-dependent processes that often involve instrumentation. Scientific observation is much more complex and sophisticated than simple sensory interaction between humans and their surroundings. According to this view, visual observation uses nothing more than crude sensors (eyes) to detect a very narrow band of electromagnetic energy. Whereas science, through the theory-guided development of technology, has produced more reliable sensors with a greater range of detection capabilities than the human eye possesses. To this extent, remote sensing observations consist of a causal chain that links a human observer to natural phenomena via the technology they employ for data collection, management and analysis (Rhoads and Thorn 1996), and that draws on background knowledge about how specific technologies can generate data on the phenomenon. This view emphasizes that there is distinction between 'data' and 'phenomena' that complements the difference between 'observation' and 'theory' (Rhoads and Thorn 1996). Remote sensing then can be construed as a search for phenomena through the acquisition of data that relies on theory-dependent technology and techniques. However, despite the use of sophisticated analytical procedures and next generation data gathering devices, the analytical perspective of remote sensing is still inclined to view observation in the traditional way. This traditional way refers to the ground/field observations. That means observation is considered to be the ideal and necessary way to verify notions about the theory driven 'search for phenomena'.

5.6 Analytical Design

Analytical design refers to the evaluation and selection of methods involved in data analysis. Analytical design is an important element of research design. The choice of how the data to be analyzed affects the choice of data/observation and the final output. Analytical methods often refer to mathematical and statistical models/algorithms. The method of understanding the relationship between visually interpreted/digitally processed information and the actual status or phenomenon on the ground and evaluating the situation is also analytical indeed.

Conventional digital analysis of remote sensing data focuses on classifying the pixels without incorporating information on the spatially adjacent data; i.e., the data is not considered as an image but as a listing of spectral measurements. There is a need to incorporate the image representation of the data in the digital analysis. Remote sensing data has the unique ability of being represented as an image. The subjective evaluations afforded by this image can be used as an interface between the human and the computer—thus supplementing numeric data with analyst experience. These inferences are developed into a procedural framework to initiate and guide the analysis of remote sensing data. Human analyst uses judgment and native intuition to make decisions while the assessment of the situational physics is a highly mathematical endeavour best left to the computer.

Madhok (2002) states that mathematical/statistical modelling on the computer serves a useful purpose, in that the system dynamics can be reduced to the manipulation of a few parameters. If applicable, the complexity of the ensuing analysis can be significantly reduced, and thus be synthesized by the researcher into a suite of analysis-routines. In contrast, the factor invaluable to the successful application of laboratory models of terrestrial phenomena is the human ability to learn and to adapt the analysis to the peculiarities of the problem. Successful analysis is thus a balance between perceptive insights and mathematics. Madhok (2002) presents a good design for remote sensing data analysis.

5.6.1 Measurement

Some scientists prefer to refer the remote sensing as *remote measurement* because the data obtained using new remote sensing systems are so accurate (Robbins 1999). Whether or not the remote sensing can be termed as remote measurement because of very high accuracy may be debated. However, it can essentially be termed as remote measurement because it measures a quantity. Whatever the indirect measurements are made in the analysis, the primary measurement of remote sensing is the amount of reflected/emitted/backscattered energy. The basic idea is simple in digital remote sensing: incoming energy is transformed into a voltage by the combination of input electrical power and sensor material that is responsive to the particular type of energy being detected. The output voltage waveform is the response of the sensor(s), and a digital quantity is obtained from each sensor by digitizing its response. Digitization involves two processes: sampling and quantization. To understand the basic idea underlying sampling and quantization, we consider one continuous image with respect to row and column coordinates, and also in amplitude. Amplitude is actually the signal strength or amount of electromagnetic energy which is responsible for the value of a pixel. Digitizing the coordinate values is called sampling and digitizing the amplitude values is called quantization. The result of sampling and quantization is a matrix of decimal numbers. Therefore, measurement is the basis of remote sensing; all further analysis is dependent on this basic measurement.

The analysis in remote sensing is based on inquiry; and analysis is driven, above all else, by the pixel values. Inquiry is often conducted within a framework of information that is relevant to the issues driving curiosity and can be used to formulate answers to research questions. Information about the world is gathered through observation. Information is essential to the production of knowledge, and observation is essential for the production of image. Therefore, knowledge, in remote sensing, comes from the image or, in specific, from pixels; i.e., from the remote measurements of electromagnetic energy. Several indirect measurements are then involved in the analysis from this basic measurement; for example, measuring the normalized differential vegetation index values for the analysis of vegetation health.

5.6.2 Variables and Relationships

Variable is a concept which can take on different quantitative values. In simple words, variable is something that changes. It changes according to different factors. Some variables change easily, like the temperature of a location, while other variables are almost constant, like the land-cover type of a built surface. Researchers are often seeking to measure variables. Variables are of two types: *independent* and *dependent*. The independent variable is the variable which the researcher would like to measure, while the dependent variable is the effect (or assumed effect), dependent on the independent variable. In case of remote sensing reflectance (pixel) value in a band is independent variable, and other (dependent) variables are derived from this independent variable. Examples of dependent variables are vegetation index values, percentage of built-up in a particular area, and so on.

Research in remote sensing primarily depends on the variation of independent variable across the geographic space, time, and electromagnetic spectrum. Pixel values vary across the spatial, spectral, and temporal dimensions. Spatial variation means values of pixels at different spatial locations are different. Spectral variation refers to the difference in values of a pixel in different bands. Temporal variation is the difference in pixel values for a given location with the change of time. These variations help us to identify and/or discriminate different types of features or their status. Therefore, the primary search of a remote sensing researcher is spatial/ spectral/temporal variations. A good deal of research in remote sensing conforms to this view, and researchers working within it often collect data across the spatial, spectral and temporal dimensions in their effort to 'explain' these variations.

The explanation relates to *causality*. Causality is the relationship between an event and a second event, where the second event is understood as a consequence of the first. The relationship or correlation is the primary consideration in causality (Wright 1921). Most explanatory science approaches causation from the viewpoint of what causes the values of an attribute or characteristic to vary from observation to observation (Visser and Jones III 2010). In this sense, the values of an attribute measured for a class or type of observation are called a *variable*; for example, reflectance of vegetation in near-infrared (NIR) band is a variable for vegetation class. Theory attempts to explain why variables vary the way they do. If, in a theory, an independent (or causal) force or factor (a) causes a dependent (or caused, or response) effect (b), then in the real world, a measurable empirical relationship should exist between the variables a and b. In particular, a and b should co-vary with one another if a causal relationship exists. Referring to the given example above, if vegetation health improves (variable a), its reflectance in NIR band (variable b) increases, or vice versa. It was mentioned earlier in Chap. 4, (Sect. 4.1) that remote sensing data analysis works on the principle of inverse theory. Therefore, if we find high reflectance in NIR band, we conclude that there is good vegetation health. If the relationship does not exist then the theory is false, although there might have been problems in the attempt to study its existence. If the relationship is shown to exist in a certain situation, then the theory is not false. Note that this is not the same as saying that the 'theory is proven true', because there may be other situations in which the relationship does not exist, or there may be a reason for the existence of the relationship other than that postulated in the theory.

What is a relationship? If observations with a certain value of an attribute a also tend to have a certain value of attribute b, then that is a relationship. Relationship may be positive or negative. If b increases with the increase of a, it is a positive relationship. For example, built-up density increases with the increase in number of working persons. A negative relationship occurs when b increases with decreased *a*, or vice versa. For example, vegetation in a city decreases with the increase in built-up. Relationships are average tendencies found in a large number of observations including remote sensing. Statistics that describe a relationship, either its closeness, nature, or both, are called *correlations*. Important to realize that, mathematically, correlation does not prove causation. If variable *a* and *b* are correlated, it is possible that *a* causes *b*, *b* causes *a*, or both *a* and *b* caused by a third variable *c* (Bernstein et al. 1988).

5.6.3 Validity and Reliability

Validity raises the question of whether or not remote sensing and other data collection devices measure the variables we think to measure. It refers to the degree in which our measuring device is truly measuring what we intended it to measure. Reliability refers to whether the measurement devices measure the variables in the same way (quantity) for each observation, or the same way each time or place it is used. Reliability is synonymous with the consistency of a test, survey, observation, or other measuring device. Therefore, validity and reliability are primarily related with observations, although the concept can be extended to justify the results from the research.

Validity and reliability in remote sensing research is involved in two instances: first in the observations and then the results from analysis. Whenever a remote sensing sensor or other measuring device is used as part of the data collection process, the validity and reliability of that data is important. This refers to the observations. Ground truth observations are often used to calibrate and validate the remote sensing observations. That means ground observations are considered to be the 'valid and ideal' (truth) to validate the reliability of remote sensing observations. Reliability, in general sense, comes from the confidence of validity. However, in remote sensing, reliability may influence the selection of data. For example, all blue, green, red and NIR bands from a remote sensing sensor are valid in respect of ground truth, however, for vegetation analysis, NIR band is more reliable than the blue band.

In general, post analysis results are also necessary to be validated. Whether one is planning a research or interpreting the findings of someone else's work, determining the impact of the results is dependent upon the validity and reliability. Essentially, validity entails the question, 'does our measurement process and assessment actually measure what we intend it to measure (whether the results relate to the ground reality)'. The reliability addresses whether repeated measurements or assessments provide a consistent result given the same initial circumstances. That means whatever the methods and models developed/used by the researcher to obtain the results, 'can it be applied repeatedly with the same confidence in similar conditions'. An important point to remember is that without the validity one cannot have reliability. Results of a poorly designed or executed research are not applicable to any future research and application; therefore, it does not add any new knowledge.

5.7 Operational Design

Operational design refers to the techniques needed for sampling design, observational design, and analytical design. Putting a research plan into action is not a straightforward task, for this is the point at which theory meets practice. Six crucial matters that all researchers must consider are as follows (Herod and Parker 2010):

- What is the purpose of the research?
- How do we understand our objects of analysis?
- How do we operationalise the research process (for example, do we use a case study approach)?
- What is the relationship between the questions we pose and the answers we generate through research?
- What implications are there of using different types of data collection?
- What value might there be in comparative approaches to research?

According to Herod and Parker (2010), these questions do not stand alone and cannot be posed or answered in a conceptual or philosophical vacuum. Indeed, although we might try to avoid issues of philosophy and simply 'get on with the research', good research cannot be conducted in a philosophical or epistemological void. In fact, given that research is a process by which we make claims about how the world is, any such claims can only make sense when interpreted through particular epistemological or philosophical lenses (refer Herod and Parker (2010) for detailed discussion).

References

Bernstein IH, Garbin CP, Teng GK (1988) Applied multivariate analysis. Springer, New York

Bhatta B (2011) Remote sensing and GIS, 2nd edn. Oxford University Press, New Delhi

Congalton RG (1988) A comparison of sampling schemes used in generating error matrices for assessing the accuracy of maps generated from remotely sensed data. Photogram Eng Remote Sens 54(5):593–600

Congalton RG, Green K (2009) Assessing the accuracy of remotely sensed data: principles and practices, 2nd edn. CRC Press, Boca Raton, FL

Congalton RG, Oderwald RG, Mead RA (1983) Assessing landsat classification accuracy using discrete multivariate statistical techniques. Photogram Eng Remote Sens 49(12):1671–1678

Fitzpatrick-Lins K (1981) Comparison of sampling procedures and data analysis for a landuse and land-cover map. Photogram Eng Remote Sens 47(3):343–351

Genderen JL, Lock BF (1977) Testing land-use map accuracy. Photogram Eng Remote Sens 43(9):1135–1137

Ginevan ME (1979) Testing land-use map accuracy: another look. Photogram Eng Remote Sens 45(10):1371–1377

Hay AM (1979) Sampling designs to test land-use map accuracy. Photogram Eng Remote Sens 45(4):529–533

Herod A, Parker KC (2010) Operational decisions. In: Gomez B, Jones JP III (eds) Research methods in Geography: a critical introduction. Blackwell Publishing, West Sussex, pp 60–76

Hord RM, Brooner W (1976) Land-use map accuracy criteria. Photogram Eng Remote Sens 42(5):671–677

Lillesand TM, Kiefer RW, Chipman JW (2004) Remote sensing and image interpretation, 5th edn. Wiley, Hoboken, NJ

Madhok V (2002) A process model for remote sensing data analysis. IEEE Trans Geosci Remote Sens 40(3):680–686

Rhoads BL, Thorn CE (1996) Observation in geomorphology. In: Rhoads BL, Thorn CE (eds) The scientific nature of geomorphology. Wiley, Chichester, pp 21–56

Rhoads BL, Wilson D (2010) Observing our world. In: Gomez B, Jones JP III (eds) Research methods in Geography: a critical introduction. Blackwell Publishing, West Sussex, pp 26–40

Robbins, J. (1999). High-tech camera sees what eye cannot. *New York Times*, Science section, September 14, D5

Rosenfield GH, Fitzpatrick-Lins K, Ling H (1982) Sampling for thematic map accuracy testing. Photogram Eng Remote Sens 48(1):131–137

Stehman S (1992) Comparison of systematic and random sampling for estimating the accuracy of maps generated from remotely sensed data. Photogram Eng Remote Sens 58(9):1343–1350

Storie CD, Bugden-Storie J (2010). Remote sensing research in undergraduate education: an international fieldwork perspective. In: Proceedings of geoscience and remote sensing symposium (IGARSS), 2010, Honolulu, HI, pp. 1114–1117. doi:10.1109/IGARSS.2010.5650049

Visser S, Jones JP III (2010) Measurement and interpretation. In: Gomez B, Jones JP III (eds) Research methods in Geography: a critical introduction. Blackwell Publishing, West Sussex, pp 41–59

Wright S (1921) Correlation and causation. J Agric Res 20(7):557–585

Chapter 6
Power, Politics, and Ethics in Research

Abstract Politics are used for the purpose of constructing and maintaining control over the research process and its products; because research produces knowledge and knowledge can be a means for obtaining the power. Ethics and politics are normally considered domains that do not mix, rather there remain conflicts. The conflict between scientific research ethics and politics is not different. There are many ways of exercising the power and politics on research and researcher; hence the role of ethics comes. Other than this, ethics is also important at personal level (personal ethics). This chapter helps to understand the nature of power and politics and the critical role of ethics in scientific research, especially remote sensing research.

Keywords Remote sensing • Scientific • Power • Politics • Ethics • Knowledge • Restriction • Plagiarism • Corruption • Intellectual honesty • Exploitation • Fraud

6.1 Power, Politics, and Scientific Research

Science is intended to generate new knowledge. The word 'knowledge' suggests certainty, authoritativeness, even usefulness. Scientific knowledge, for example, is what the bulk of relevant specialists agree on at any particular time. It is a good thing to be knowledgeable. However, knowledge can be biased in various ways, for example by providing a restricted picture of economic behaviour. Researchers are often forced to do a specific thing rather than driving by whim or curiosity. An excellent example can be given from Martin (1998). If a pharmaceutical company sponsors a research into drugs to reduce tension or control hyperactivity, then that is what the researchers are likely to find if they are successful. Funding alone does not guarantee the results, of course, but if something is found it is likely to be of more value to the funder than others. The drug researchers might, in the course of their investigations, happen upon a substance that does something different, such as preventing kidney stones. But they are unlikely to do much research on unpatentable substances or methods, since there is no profit in that. They certainly would not find a way to reduce tension that does not involve drugs at all, such as by relaxation, biofeedback or small group dynamics, since they are looking only

B. Bhatta, *Research Methods in Remote Sensing*, SpringerBriefs in Earth Sciences, DOI: 10.1007/978-94-007-6594-8_6, © The Author(s) 2013

at drugs. Funding, then, provides a strong steering process. Only certain types of knowledge are likely to result because the researchers are paid to look only for certain types of things.

Funding for the majority of formal research in the world today is provided by governments and corporations. The amount of funding from trade unions, environmental groups or women's groups is tiny by comparison. That means most research follows governmental or corporate agendas (Martin 1998; Dickson 1984; Newson and Buchbinder 1988; Ridgeway 1968). Having the steering in the government and corporate hands, science has become a field of battle. Corporate culture is how to make more profit; it is perhaps not unethical to some extent. But government's role is surely unethical; it is public money that government offers to the researchers. However, the politics of science is nothing new. Galileo, for example, committed a political act in 1610s when he simply forced to write against of what he proved. This writing proved that Copernicus had been right in 1543: sun revolves around the earth, not the other way around, as contemporary opinion and the Roman Catholic Church held. The same tradition is being continued. For example, in 2005, leaked documents revealed that the chief of staff for the White House Council on Environmental Quality, a former oil industry lawyer, had altered climate reports to soften scientific findings showing that fossil-fuel use and deforestation triggered global warming (Vergano 2007). Vergano (2007) documented many such examples.

Otto (2011) stated that science is inherently political and that the practice of science is a political act. "Knowledge and power go hand in hand", said Francis Bacon, "so that the way to increase in power is to increase in knowledge". Otto (2011) explained the relationship between knowledge and power as: "At its core, science is a reliable method for creating knowledge, and thus power. Because science pushes the boundaries of knowledge, it pushes us to constantly refine our ethics and morality, and that is always political. But beyond that, science constantly disrupts hierarchical power structures and vested interests in a long drive to give knowledge, and thus power, to the individual, and that process is also political". Power tends to corrupt, and knowledge power is no exception. Intellectuals on their own are not major wielders of power. They mostly operate to serve other powerful groups, especially governments, corporations and professions, by providing useful knowledge and by providing legitimacy for policies and practices (Derber et al. 1990).

Some scientists view the science as above of politics and power. They are, perhaps, the assets of science and society; alas, they are bound to limit their activity just because of 'fund'. Some academics argue that they should be given full academic freedom, without constraints from government and corporate funders. But Martin (1998) stated that this is really just a claim for funding without accountability. He has proposed an alternative model of research—community participation and control. Community participation means that anyone potentially could join in research projects: no credentials would be required. Community control means that funding and accountability would be in the community's hands. Whether this model can solve the problem or not may be debated because the 'control' does not disappear.

Another important thing is politics of publication. Scientific research has been overtaken by careerism and a management culture, to the detriment of originality and discovery. That's the view aired by Lawrence (2003). He argues that scientists, desperate to publish in a few top journals, are wasting time and energy manipulating their manuscripts and courting editors. This issue will be further elaborated in Sect. 6.2.

6.1.1   Power, Politics, and Remote Sensing

Remote sensing has a very long history dating back to the end of the 19th century when cameras were first made airborne using balloons and kites. The advent of aircraft further enhanced the opportunities to take photographs from the air. Then satellite mounted sensors had been developed to operate it from the space. Whatever the developments we see in the field of remote sensing were primarily for the military (for the power and politics). It was completely driven by power and politics. Remote sensing had been nourished within the core of power and politics. Initially it was not available to the civilian researchers. Most of the significant developments in remote sensing came just for World War-I and II.

Although, many civilian remote sensing satellites have been launched since 1972 (starting from Landsat-I), still spy satellites, nano satellites, and high resolution sensors are being launched by the governments for power and politics. Very high resolution images are still restricted to the civilians in many countries including United States. For example, GeoEye-1 captures imagery at a spatial resolution of 0.41 m; however, it is downsampled to 0.5 m for the civilians because the US government does not allow higher than 0.5 m resolution to the civilians. Arial photography is still performed only by the governments in many countries and photographs are restricted; for example, India. In India (and many other countries), an individual researcher is not entitled to purchase even a low resolution satellite image. She/he needs to be associated with some institution and some sort of declaration is mandatory by the head of institution to obtain the imagery. That means, as an individual, one cannot perform the research with their own fund.

This type of restrictions is everywhere and was always there. On December 3, 1986, the United Nations had faced the difficulty to pass "Principles Relating to Remote Sensing of the Earth from Outer Space". The United States' position had been that collection and distribution of civilian remote sensing imagery should be unrestricted. The Soviet Union's position was to ensure that acquisition and distribution of imagery should only be allowed with consent of the state that is overflown. It was the case for the outer space; if it comes within the air? Aerial remote sensing, till date, cannot be performed beyond the political boundary. Matthew (1983) is an essentially referred text in this context. Whether these political restrictions are good or bad is not the issue of this book. The issue is, rather, these political restrictions have created a knowledge gap in civilian remote sensing research and applications.

Why are these restrictions imposed? India restricts its residents (individuals) to the access of very high resolution images whereas Pakistani (or any other) military can purchase one such image covering India from a commercial vendor (from other country) without having any problem. Does it make any sense? Does it suggest rethinking on the data policy? Whatever the answer may be, politics does not want to make everything free, especially the remote sensing data of having high value and importance. It can be seen as an internal conflict. A good example of this internal conflict is the politics of remote sensing capabilities. No country wants to be left behind, but on the other hand, why should countries expend scarce resources acquiring the launch vehicles, satellites, and infrastructure needed to support a remote sensing program when much of the end product (images, etc.) can be purchased at a modest cost from commercial vendors. Remote sensing technology provides countries with the ability to evaluate others' capabilities to a degree that is totally unprecedented in the history of relations between countries. The countries that employ this technology can assess others' military and—to some degree—economic capabilities (refer Ammons 2010). It also has the effect of lessening the deception possible by a closed society in concealing its capabilities. This technology, in another way, could be said to have the potential to stabilize the international system.

Now, the question is whether this technology actually makes a country more secure or if it increases the perception, both internally and externally, that it is more secure. Perhaps it is a little of both. Countries will always seek more information about their adversaries and any technology that will increase the quantity and quality of that information is valuable because of its real or perceived contribution to the country's security. Most countries that own remote sensing technology profess to employ this technology for peaceful purposes. It is difficult to argue that activities such as resource management and disaster management are anything other than positive pursuits. However, it would be a simplified thinking to assume that a country concerned about its security (all countries are concerned about security) would not employ every available means to protect itself (Ammons 2010). This is particularly true if these means are defensive rather than offensive and can be accomplished with some measure of privacy. However, we must assume all countries that own remote sensing technology gather imagery intelligence of other countries' military capabilities. This information are collected mainly for three reasons—firstly, to monitor whether a country is violating any international agreement (defensive in nature); secondly, to prepare its own military capabilities to that standard (or higher) of other countries (defensive in nature); and thirdly, to use this information during a war or to attack a country (offensive in nature). The war among countries, perhaps, will not be stopped ever. Therefore, offensive use of remote sensing will also be continued for ever.

6.2 Ethics in Research

The politics of research merge with its ethics. Both are concerned with the normative, and areas of overlap are obvious. Here we concentrate on questions of what is good (or bad) research from an ethical or moral point of view, recognizing that there is an element of politics in some of this. It is important to stress at the outset the distinction between what is good (or bad) research from an ethical as opposed to scientific/technical point of view. The scientific research enterprise is built on a foundation of trust. Scientists trust that the results reported by others are valid. Society trusts that the results of research reflect an honest attempt by scientists to describe the world accurately and without bias. But this trust will endure only if the scientific community devotes itself to exemplifying and transmitting the values associated with ethical scientific conduct (NAS 2009).

What comes under research ethics? The answer is not very straight. Different literature raised different issues. One may find lot of articles online address-ing these issues. However, some things are surely wrong from an ethical point of view; let us begin with things that are surely wrong. It is wrong to *tamper the data* (also known as *falsification of data*), by inventing it or changing actual data to make a case more persuasive. It is also wrong to *tamper the results*, like a sta-tistical test, to suggest a better fit to some hypothesis than is actually the case. It is very important to keep in mind that science is performed by humans, with the same foibles as other people. It is a different issue that scientists sometimes make mistakes. But, scientists also sometimes lie, just like others. Scientific fraud is rel-atively rare, though, because if one is found to have committed fraud as a scien-tist, her/his career is really weakened, if not completely finished. Sooner or later, someone somewhere might well try to repeat one's work and find out they cannot; such cross-checking is a great deterrent. Checking and rechecking is continuously performed in science; that is why the scientific method is so powerful. It is built to handle error and deceit and to be (sometimes very slowly) self-correcting. Because of this, the scientific method is the best way we now know of to get near the 'truth' about the world around us (Anderson 2000). Fraud and falsification of data are so inimical to scientific method that almost never do scientists succumb to their lure of quick rewards. Even a single case of scientific fraud, when publicized, does unimaginable damage to the credibility of scientists in general, for the public can-not confirm our findings; they must trust them. Fraud also slows the advance of a scientific field (for experiments that rarely are exactly replicated) and fraud is not suspected until all alternative explanations have been eliminated (Jarrard 2001).

Next very serious issue is *plagiarism*. It is the 'presentation' of the work of another person as one's own or without proper acknowledgement. Plagiarism violets copyright and intellectual property rights. The Concordia University of Canada has explained this very clearly as follows: "This could be material copied word for word from books, journals, internet sites, professor's course notes, etc. It could be material that is paraphrased but closely resembles the original source. It could be the work of a fellow student, for example, an answer on a quiz, data for a lab report, a paper or assignment completed by another student. It might be a paper purchased through one of the many available sources. Plagiarism does not refer to words alone—it can also refer to copying images, graphs, tables, and ideas. 'Presentation' is not limited to written work. It also includes oral presenta-tions, computer assignments and artistic works" (refer http://www.concordia.ca/programs-and-courses/academic-integrity/plagiarism/). If one uses her/his own work (that has been published or delivered/sold) without the correct citation, this too is plagiarism. It is not plagiarism, however, to use other writers' material when one acknowledges whose material it is. However, plagiarism, as explained above, is a part of honest research writing. Plagiarism, as associated with research method, is to deny others credit for their research findings, say by failing to acknowledge them or properly reference their work or deliberately stealing their results/research work. In case of collaborative research, co-researchers must be given proper credit. Intellectual plagiarism, the attempt to take credit for the ideas of others, is clearly unacceptable, but its boundaries are indefinite. Jarrard (2001) stated that intellectual plagiarism is more often suspected than deliberately practiced. Ideas frequently stem from interactions with others. In such cases, the combination of two perspectives deserves credit for development of the idea, not the person who first verbalizes it. Perhaps the idea is not even verbalized during the discussion, yet one of the individuals later 'realizes' the idea when solitarily thinking about the subject.

Whoever undertakes research comes to the task with what is referred to as *positionality* (Rose 1997). In other words, they may be influenced in one way or another by personal characteristics such as social origin, race, and gender. In terms of research, these could be an undue advantage as well as a disadvantage. Any research may invade the *privacy*. It is applicable for a team rather than individual. A research team should always maintain privacy regarding the research subject. Nothing should be disclosed or published without the consent of other researchers or the organization (for which the research is being conducted). Then there is the issue of *confidentiality*, which may require disguising the respondents and the location of the project. Finally, if the research is published (and there are ethical reasons which it should be, so as to make the findings widely available), those who provided data might be given copies so that they can see the use to which their information has been put, as well as to provide reassurance if confidentiality has been an issue (Smith 2010).

An important geographical question that raises political as well as ethical issues is where to conduct remote sensing research. The point just raised, about whether local residents might gain or lose from research, has implications for work in the impoverished inner city, for example (Smith 2010). Smith (2010) explained this concept very clearly—"It could be asserted that there is a potentially exploitative relationship between a university and its local (often poor) neighbourhood, if the area is frequently used to provide research subjects who gain nothing from the process. This underlies the ethical aspects of expeditions and similar arrangements". At a wider scale, it is common for geographers from the United States and Western European countries to undertake research in the underdeveloped world (Sidaway 1992), with the risk of similar exploitative relationships if nothing is offered in return. At the least, it is sometimes asserted, those working in such places should involve local geographers in a collaborative role, to assist with capacity building and otherwise pass on some of the benefits of coming from well-endowed universities in affluent countries where conditions are the envy of many others elsewhere. "This is part of the wider ethical issue of what we in the more fortunate parts of the world may owe to 'distant strangers'" (Smith 2010). One may refer Smith (1988, 1994, 2003, 2004) for further detail on this issue.

6.3   Corruptions of Expert Knowledge

Some people say "knowledge is power". If knowledge is power and power corrupts, does knowledge corrupt too? Martin (1998) stated that knowledge is not power just by itself, but it can be a means for obtaining power, wealth and status. Because of this, individuals and groups try to convince others that they have exclusive access to the truth—in other words, that they are the authorities in particular areas of knowledge. In order to part with this knowledge, they ask for fees, honour, career and status. Because there can be money and status from being a recognized expert, there is a temptation for experts to sell themselves to the highest bidder. Many experts are willing to serve those who are powerful, who are not necessarily those who need expert knowledge the most.

Once a group of experts has established itself as having exclusive control over a body of knowledge, it is to their advantage to exclude nonexperts. This occurs in many ways. A long and expensive training is commonly demanded before a new-comer can be accepted as an expert. In the case of medicine, law, engineering and some other professions, formal certification is required in order to practise in the field. The new recruit is expected to use the appropriate jargon. Editors expect a certain approach and type of writing for contributions to expert journals. Many editors shamelessly ask the authors to refer papers from their journals, just to get higher index value. Some editors/reviewers of society journals give preference to the members of the society (or well known experts); although their papers may not be of very high quality. Whereas, high quality papers from non-members/nonexperts are criticized.

Nonexperts (e.g., research scholars), who are working under an expert, are often exploited in various ways; even they are sometimes treated as paid servants. There are also some instances where a share of the scholarship money of junior researcher is taken by the research guide. Female junior researchers often become victim of sexual harassment in exchange of knowledge by these 'experts'. These types of exploitation are more serious in developing or conservative countries.

Martin (1998) stated that many experts are arrogant, displaying contempt or hostility to amateur interlopers. Knowledge experts are not inherently nasty. Rather, the power they gain from having control over the field leads them to develop attitudes, beliefs, training systems and procedures that maintain the control and keep out nonexperts. The conflict is between the expert establishment, namely the group of experts with official recognition and more power, versus expert outsiders (Martin 1996). Even more serious is when an expert who is part of the establishment becomes a dissident, questioning the standard way of doing things. These researchers are harassed, ostracised, reprimanded, demoted and dismissed. Instead of responding to the person by discussing the issues and attempting to refute their views, the dissident becomes the target. This can only happen when the establishment has power that can be exercised against dissidents.

6.4 Personal and Professional Ethics

Personal and professional ethics are not distinguishable; all ethics are personal
(Jarrard 2001). A scientist must make ethical decisions with care, not only because
they affect self image but also because, as Sindermann (1987) has pointed out, scientific reputations are hampered. Some rules of scientific ethics are universal, and
others are subjective. All require personal judgment. Not all of the ethical opinions that follow can claim consensus. "Scientists are not democratic; most insist
on deciding personally whether a rule warrants following, rather than accepting
the majority vote" (Jarrard 2001).

   Therefore, every ethical decision must be weighed personally and subjectively.
Before making a final decision on any ethical issue, it is worthwhile to consider
the issue from the standpoint of Kohlberg's (1981, 1984) criterion for mature
moral judgment: does the judgment hold regardless of which position one occu-
pies in the conflict? Kohlberg's criterion sounds almost like a generalization of
"Do unto others as you would have them do unto you". The habit of applying
Kohlberg's criterion is analogous to the habit (or skill) of objectively evaluating
the effect of data on various hypotheses, without regard for which hypothesis one
favours (Kuhn et al. 1988).

   Intellectual honesty must be a goal of every scientist (Jarrard 2001). People tend
to ignore evidence that diverges from their expectations. A researcher must fight
against this tendency; continued awareness and evaluation of possible personal biases
is the best weapon. Intellectual honesty requires that we remain alert to conflicts of
interest, whenever we review proposals and manuscripts, and wherever objectivity
and personal advancement clash. "Intellectual honesty requires that we face weaknesses as well as strengths of data, hypotheses, and interpretations, without regard for
their origin, invested effort, or potential impact on our beliefs" (Jarrard 2001).

6.5 Ethics in Remote Sensing Research

As stated by Slonecker et al. (1998): "Fundamental changes are taking place in the
world of remote sensing with respect to three primary developments. First, a new
generation of space-borne sensors will be able to deliver high spatial and spectral resolution imagery on a global basis. Technical advances are making previous
restrictions on data scale, resolution, location, and availability largely irrelevant.
Second, economic restructuring of the remote sensing community will transform
the control and distribution of imagery and imagery-derived information generally away from government and into the private sector. Third, the development
of a digital, global information infrastructure, such as the Internet, will allow
for rapid global distribution of information to a worldwide user community. The
combined effects of these developments could have significant legal and ethical
consequences for all remote sensing professionals".

The American Society for Photogrammetry and Remote Sensing maintains a code of ethics for their members (refer http://www.asprs.org/a/membership/certification/appendix_a.html). According to them, "honesty, justice, and courtesy form a moral philosophy which, associated with mutual interest among people, should be the principles on which ethics are founded". The code of ethics by the American Society for Photogrammetry and Remote Sensing would be a good reference in this context: Each person in the mapping sciences (Photogrammetry, Remote Sensing, Geographic Information Systems, and related disciplines) profession shall have full regard for achieving excellence in the practice of the profession and the essentiality of maintaining the highest standards of ethical conduct in responsibilities and work for an employer, all clients, colleagues and associates, and society at large, and shall

1. Be guided in all professional activities by the highest standards and be a faithful trustee or agent in all matters for each client or employer.
2. At all times function in such a manner as will bring credit and dignity to the mapping sciences profession.
3. Not compete unfairly with anyone who is engaged in the mapping sciences profession by:
   a. Advertising in a self-laudatory manner;
   b. Monetarily exploiting one's own or another's employment position;
   c. Publicly criticizing other persons working in or having an interest in the mapping sciences;
   d. Exercising undue influence or pressure, or soliciting favours through offering monetary inducements.
4. Work to strengthen the profession of mapping sciences by:
   a. Personal effort directed toward improving personal skills and knowledge;
   b. Interchange of information and experience with other persons interested in and using a mapping science, with other professions, and with students and the public;
   c. Seeking to provide opportunities for professional development and advancement of persons working under his or her supervision;
   d. Promoting the principle of appropriate compensation for work done by person in their employ.
5. Undertake only such assignments in the use of mapping sciences for which one is qualified by education, training, and experience, and employ or advise the employment of experts and specialists when and whenever clients' or employers' interests will be best served thereby.
6. Give appropriate credit to other persons and/or firms for their professional contributions.
7. Recognize the proprietary, privacy, legal and ethical interests and rights of others. This not only refers to the adoption of these principles in the general conduct of business and professional activities, but also as they relate specifically to the appropriate and honest application of photogrammetry, remote sensing, geographic information systems, and related spatial technologies. Subscribers to this code

shall not condone, promote, advocate, or tolerate any organization's or individual's use of these technologies in a manner that knowingly contributes to:

a. deception through data alteration;
b. circumvention of the law;
c. transgression of reasonable and legitimate expectation of privacy.

These codes of ethics essentially provide some guidelines, but not all. Ultimately, the researcher needs to decide on 'good' or 'bad' at every step in her/his research. Therefore, the ethics must come from the inner soul of the researcher. One should remember that 'ethics is the ultimate power'.

References

Ammons AA (2010) Competition among states: case studies in the political role of remote sensing capabilities. PhD Dissertation, Catholic University of America, Washington. URL: http://aladinrc.wrlc.org//handle/1961/9175

Anderson G (2000) Scientific method. Online lecture note. URL: http://web.archive.org/web/20060217052458/http://pasadena.wr.usgs.gov/office/ganderson/es10/lectures/lecture01/lecture01.html

Derber C, Schwartz WA, Magrass Y (1990) Power in the highest degree: professionals and the rise of a new mandarin order. Oxford University Press, New York

Dickson D (1984) The new politics of science. Pantheon, New York

Jarrard RD (2001) Scientific methods. Online book, URL: http://emotionalcompetency.com/sci/booktoc.html

Kohlberg L (1981) Essays on moral development: the philosophy of moral development vol 1. Harper and Row, New York

Kohlberg L (1984) Essays on moral development: the psychology of moral development vol 2. Harper and Row, New York

Kuhn D, Amsel E, O'Loughlin M (1988) The development of scientific thinking skills. Academic Press, San Diego

Lawrence PA (2003) The politics of publication. Nature 422:259–261. doi:10.1038/422259a

Martin B (ed.) (1996) Confronting the experts. State University of New York Press, Albany

Martin B (1998) Information liberation: challenging the corruptions of information power. Freedom Press, London

Matthew M (1983) The technical, legal and political implications of remote sensing satellites. Theses and dissertations (Comprehensive), Paper 54. http://scholars.wlu.ca/etd/54

NAS (National Academy of Sciences) (2009) On being a scientist, 3rd edn. The National Academies Press, Washington, DC. Available at: http://www.nap.edu/catalog.php?record_id=12192

Newson J, Buchbinder H (1988) The university means business: universities, corporations and academic work. Garamond Press, Toronto

Otto SL (2011) Good science always has political ramifications. scientific american, 24 Nov 2011. URL: http://www.scientificamerican.com/article.cfm?id=good-science-always-has-political

Ridgeway J (1968) The closed corporation: american universities in crisis. Random House, New York

Rose G (1997) Situating knowledge: positionality, reflexivities and other tactics. Area 26:305–320

Sidaway JD (1992) In other worlds: on the politics of 'first world' geographers in the 'third world'. Area 24:403–408

Sindermann CJ (1987) Survival strategies for new scientists. Plenum Press, New York

Slonecker ET, Shaw DM, Lillesand TM (1998) Emerging legal and ethical issues in advanced remote sensing technology. Photogram Eng Remote Sens 64(6):589–595

Smith DM (1988) Academic links with South Africa: is ignorance a greater sin? Area 20:357–359

Smith DM (1994) On professional responsibility to distant others. Area 26:359–367

Smith DM (2003) Geographers, ethics and social concern. In: Johnston RJ, Williams M (eds) A century of british geography. Oxford University Press, Oxford, pp 625–644

Smith DM (2004) Morality, ethics and social justice. In: Cloke P, Crang P, Goodwin M (eds) Envisioning human geographies. Arnold, London, pp 195–209

Smith DM (2010) The politics and ethics of research. In: Gomez B, Jones JP III (eds) Research methods in geography: a critical introduction. Blackwell Publishing, West Sussex, pp 411–423

Vergano D (2007) Science vs. politics gets down and dirty. USA Today, 7 Aug 2007. URL: http://usatoday30.usatoday.com/news/washington/2007-08-05-science-politics_N.htm

Chapter 7
Documenting the Research

Abstract Any systematic investigation towards increasing the sum of knowledge can be termed as research. Every piece of research must make an original contribution to knowledge, irrespective of the method of enquiry. At the end of our research there remains the task of organizing the results and making them known to others. Our research would hardly be of any value unless we disseminate our findings to others working or interested in the same sphere of activity or knowledge. This chapter is aimed to discuss the methods and issues involved in documenting a research outcome. All three ways of documenting the research results—research paper, dissertation, and thesis—will be addressed in separate sections. Referencing styles will also be discussed in context. Finally, an attempt will be made to present a general guidance on writing scientific literature.

Keywords Research • Research   paper • Dissertation • Thesis • Writing • Referencing • APA • Style

7.1 Introduction

Communication of results, particularly via publication, is an essential part of a scientist's life. We can present our findings in a professional gathering, or if we wish to reach a wider audience, we can publish our work in a journal. Research results from academic courses are documented as dissertation and thesis. In every case, we shall have to organize our facts and opinions so as to present their meaning as logically, lucidly and tellingly as possible. Research results can be documented in three different ways: *research paper*, *dissertation*, and *thesis*. All these research documents are basically academic prose; however, some differences also exist.

Research paper (also called *academic paper, scientific paper, investigative paper, library paper*, or *scholarly paper*) is an organized academic prose published in academic journals or in proceedings of professional gatherings (e.g., seminar, conference, workshop, etc.). It contains original research results or critical review of existing results or researches. Such an organized analysis of a subject is written mainly to record and disseminate information or knowledge, or to present a point

B. Bhatta, *Research Methods in Remote Sensing*, SpringerBriefs in Earth Sciences,
DOI: 10.1007/978-94-007-6594-8_7, © The Author(s) 2013

of view on a selected topic. In fact, it is a long essay often supported by relevant references from suitable sources. It gives a succinct account of the work done, the materials used, the methods adopted, and the results arrived at. A research paper must be the first disclosure containing information to enable peers (1) to assess observations, (2) to repeat experiments, and (3) to evaluate intellectual processes (Raman and Sharma 2004). When the findings of our research are published in the form of a research paper, our efforts acquire a permanent value. However, it is worth noting that all research journals of repute control the number of papers published by a rigorous referring system to ensure the originality and quality of contribution.

A thesis or dissertation is a scholarly written document by a student of higher education, which is a required output for an academic degree. One of the differences between a research paper and a thesis/dissertation is the purpose as a thesis/dissertation is a document written in support of obtaining an academic degree or qualification. It is usually longer than research paper and thus completed in a number of years. A thesis/dissertation is usually associated with postgraduate studies, i.e. research of taught master's degree, PhD or MPhil level and is carried out under the supervision of a professor or an academic of the university. A research paper can be written by one or more author(s); however, a thesis/dissertation is always an individual submission.

The words 'dissertation' and 'thesis' often used interchangeably, there are similarities and differences though. Dissertation is the same as a thesis in many institutions and countries. If one checks the dictionary, the two are defined the same way. However, in some countries, these two are different. Importantly, they are categorized differently in different countries and institutions. A dissertation is examined at the master's level and a thesis in the doctoral level while in others the two are interchanged. In some countries, these two are labelled as *master's thesis* and *doctoral thesis*. For our discussion, let us consider dissertation for master's level and thesis for doctoral level.

A dissertation is longer than a research paper and it arises out of the study, research, and analysis undertaken by a student over a semester or a term. Hence it is also known as *term paper*. A thesis is longer and more detailed than a dissertation. It may span over, in general, a period of two to five years. In a few cases it may extend beyond five years also. One may present extensive research on a particular topic in the form of a thorough analysis, supported adequately by statistical data, survey findings, experimental results, and the like.

Although these three forms of written communication (research paper, dissertation, and thesis) vary in length, they exhibit a lot of similarities in terms of their characteristics. The following discussion will enable one to understand them in terms of their characteristics and structural elements.

7.2 Research Paper

In its style, structure, and approach, a research paper closely resembles a formal report. In certain respects, it differs from a formal report. Given below are the main characteristics of a research paper.

1. A research paper is a most important form of expository discourse. It may be written on any topic or subject, e.g., scientific, technical, social, cultural, etc., but the treatment is scholarly in nature.
2. It is highly stylized and contains a high concentration of certain writing techniques such as definition, classification, interpretation, abstraction, description, etc.
3. It is objective in nature and the presentation of information is accurate, concise, direct, and unambiguous.
4. Most research papers are characterized by the use of graphic aids, and scientific, technical, or specialized vocabulary.
5. Every research paper is a unified composition arising out of the study of a particular subject, assembling the relevant data, and organizing and analyzing the same.
6. A research paper is a documented prose work. All important analyses have to be supported by adequate evidence. In short, documentation is essential for all research papers.

A research paper is a piece of written communication organized to meet the needs of valid publication. It is therefore highly structured with distinctive and clearly evident component parts. Most general parts of a research paper can be listed as: title, authors and addresses, abstract, keywords, introduction, materials and methods, results, discussion, conclusions, references.

7.2.1 Title

The title of a research paper may be defined as the fewest possible words that adequately describe the contents of the paper. It ought to be well studied and should give a definite and concise indication of what is to come. In preparing the title for a paper, the author would do well to remember that the title will attract a person to read the full paper. Therefore, all words in the title should be chosen with great care and their association with one another must be carefully managed. It is worth remembering that the indexing and abstracting services depend heavily on the accuracy of the title. Also, an improperly titled paper may be virtually lost and may never reach its intended audience.

7.2.2 Authors and Addresses

Unless the writer of a research paper wishes to publish anonymously, a full name and a full address should be considered obligatory. The authors' names should be spelt and given in the same way in all their publications. Departure from this causes confusion at the time of accumulation of information. The listing of authors should include only those who actively contributed to the overall design and execution of the experiments. Further, the authors should normally be listed in order of importance to the experiments, the first author being acknowledged as the senior author and primary progenitor of the work being reported. An address serves two purposes: to identify the author and to supply the author's mailing address. If different authors of the same paper have different addresses, the respective names and addresses should be given separately.

7.2.3 Abstract

With increased importance acquired by secondary services, particularly the abstracting periodicals, the abstract of a research paper has assumed special significance. It has two main functions: (1) to enable readers identify the basic content of a document quickly and accurately in order to determine its relevance to their interests and thus to decide whether they need to read the document in its entirety, and (2) to meet the requirement of the abstracting journals.

There are two types of abstracts: *informative* and *indicative*. Normally a research paper should have an informative abstract which gives information about the purpose of the study, newly observed facts, conclusions of an experiment or argument, and, if possible, the essential parts of any new theory, treatment, apparatus, technique, etc. The abstract supplants the need for reading the full paper and hence should be self-contained with regard to the new information in the paper. The other common type of abstract is indicative abstract (descriptive abstract). This is more suitable for long, descriptive papers or review articles. An indicative abstract indicates the contents of the paper and the scope of the work done without giving information about the results and conclusions.

The characteristics of an abstract are (Raman and Sharma 2004): (1) should be as concise as possible and should not exceed 3 % of the total length of the paper; (2) self-contained; (3) does not contain any bibliography, figure, or table references; (4) does not contain any obscure abbreviations and acronyms; (5) generally written after the paper is prepared. The steps involved in preparing an abstract are as follows: (1) read the introductory paragraph of the study to identify objective; (2) scan the summary and conclusions at the end for noting the main findings of the study; (3) read through the text for information on methodology adopted, new data, and any other vital information; (4) prepare a draft arranging the various items in the order: objective, new methodology or equipment employed, data of fundamental value, and major conclusion and/or correlations derived; (5) modify and trim it to get the required size.

7.2.4 Keywords

A keyword is an index entry that identifies a specific research paper. Most electronic search engines, databases, or journal websites will use the words to decide whether and when to display your paper to interested readers. Journals, search engines, and indexing and abstracting services classify papers using keywords. Thus, an accurate list of keywords will ensure correct indexing and help showcase one's research to interested groups. This in turn will increase the chances of one's paper being cited.

Here is how one can go about choosing the right keywords for a research paper: (1) read the paper thoroughly and list down the terms/phrases that are used repeatedly in the text; (2) ensure that this list includes all main key terms/phrases and a few additional key phrases; (3) include variants of a term/phrase (e.g., spatial metric and landscape metric), procedures, etc.; (4) include common abbreviations of terms (e.g., GIS, GPS); (5) now, refer to a common vocabulary/term list or indexing standard in the intended discipline (e.g., GeoRef, Scopus, etc.) and ensure that the terms that have been used match those used in these resources; (6) finally, before submitting the article, type the keywords into a search engine and check if the results that show up match the subject of the paper. This will help an author to determine whether the keywords are appropriate for the topic of a paper.

7.2.5 Introduction

The purpose of an introduction is to supply sufficient background information so as to allow the reader to understand and evaluate the results of the study. It may therefore become necessary to refer to work done earlier only in strict relevance to the above purpose. Sometimes it is necessary to outline the author's earlier attempts to solve the problem along with citations to relevant literature. It is, however, redundant to attempt a complete historical survey of the earlier work. Very often it is possible to cite a single reference to an important recent review article instead of giving a long list of references; all of them might have been referred to in the review article.

Guidelines for a good introduction are as follows: (1) it should present, with all possible clarity, the nature and scope of the problem investigated; (2) should review the pertinent literature to orient the reader; (3) should state the method of investigation and, if necessary, the reasons for the choice of a particular method; (4) should state the research objectives.

7.2.6 Materials and Methods

The main purpose of this section is to describe (and if necessary defend) the data, experimental design, experimental technique, or theoretical derivation, and then provide enough details so that a competent worker can repeat the experiments. If a well-known technique or approach is used, it is enough to cite the relevant literature reference where the description is available. There is, of course, a need to depart from this policy in those cases where the original source is an obscure one. In such cases where the technique or approach adopted involves some modification over the earlier technique or approach, detailed description of the modification only need be given. For materials and input data, relevant specifications must be given. Experiments performed, the ranges covered, the new equipment used, etc. must be described in sufficient detail. Besides technical specifications, justification of preference to them must also be included. This section may have several subheadings.

7.2.7 Results

This section forms the core of the paper—the outcome. There are three ways of presenting the results: (1) in text; (2) in tabular form; and (3) in illustration form. A particular set of results should be given only in one of these forms. Duplication should be avoided as far as possible. While selecting resulting data for inclusion into a paper, one should avoid the two extremes in this respect. One extreme is the tendency to shift almost the entire data present in the laboratory notebook to the paper. At the other extreme is the situation where data is so insufficient that the reader cannot even understand the logic of the conclusions drawn. The ideal thing to do is to give such essential data as form the basis for the major conclusions emerging from the study.

Simple and descriptive data should be given in the text. Only such data as deserve to be highlighted should be given in tabular or graphical form. Information given in tables or figures should supplement, not duplicate, the information provided in the text. As to the choice between tabular and illustrative forms of presentation, the broad criterion is that for values which are given for their data value (need for high degree of exactness), tabular presentation is preferable. When one wants to highlight trends, prefer graphical form.

Tables should not be made complex by including too many items and too many details about them. Rather than making a cumbersome single table, complex data should be divided and presented in two or more tables. In the case of big tables, which can neither be shortened nor be split up, the practice of printing on a bigger sheet and folding the same should be avoided. After some years, through ageing of paper and constant use in the library, the paper is likely to crack at the fold. Therefore, the table must be composed in multiple pages. Of course, the tide and the column headings have to be repeated on the subsequent pages.

In case of illustrated form, one must remember, each figure adds to the cost of production. Therefore, only such figures should be included which are absolutely essential. Considerable economy in cost and space can be achieved in a number of ways, such as (1) combining several simple graphs into a composite illustration, particularly when either both the parameters or only one is common, and (2) making judicious alterations in scales to reduce the size of the illustrations. For economy of using space, efforts should be made to accommodate as many illustrations as possible in column size; page width size should be chosen only in essential cases.

All tables and figures should be referred to in the text as 'Table 1', 'Fig. 1', 'Chart 1', etc. and not by expressions like 'below', 'above', 'preceding', or 'following'. Captions and legends should be simple, but self-explanatory. Inclusion of too many explanatory notes and other details inside illustrations should be avoided. Line diagrams must be of at least 200 dots per inch, and pictures/photographs must be of at least 300 dots per inch.

7.2.8 Discussion

The main functions of this section are to interpret results and to highlight the significant features of the research outcome and the possible causes of these features. It should also mention the limitations, if any, of the results/methods and point out any sources of error. The tendency to repeat description of data in this section should be avoided. The strengths of the methods and results should also be discussed.

7.2.9 Conclusions

Here we must sum up our findings. One must be careful not to repeat the exact words that have been already used; we nevertheless must, in a sense, say what we already have said. In the conclusion, an academic reader will expect to see our findings summarized, and our main points spelled out clearly and concisely. In such cases where the study has led to clear-cut findings, it is preferable to give the conclusions in the form of a series of numbered points. But if the conclusion stops as a summary, the author is missing an opportunity to show how original and important the work is. The best conclusions do summarize, but they also move at least one step farther, suggesting, for example, the broader implications of the work in question. One of the major functions of conclusions is to make recommendations based on the results of the study. Be careful, however, to avoid dropping entirely new ideas and entirely new evidence into the conclusion section. The conclusion is not the place to build a new argument—it is the place to wind up the old one.

7.2.10 References

The main purpose in citing references to the work of earlier researchers is to enable the reader to consult the original source. Therefore, unless the references are complete in respect of all bibliographic details, the readers will face immense difficulty in locating the original sources. Only such references should be cited as have been actually consulted. References taken from second hand sources should not be cited. Unpublished papers and personal communications should not be listed under the reference, but should be mentioned in the text. There are several styles exists for referencing. Please refer Sect. 7.5 for discussion on referencing style.

7.3 Dissertation

A dissertation (for master's degree) does not normally exceed 20,000 words (different universities may have different guidelines) and is presented on the candidate's work and its development, its cultural, historical, and theoretical references. The dissertation should: (1) concern itself with the intellectual, visual, or cultural context of a candidate's work and the development of that work; (2) be a critical statement and not a mere exposition; and (3) identify and discuss the work's references (cultural, historical, and theoretical).

The essential features of a dissertation are as follows: original; demonstrative; extensive, relevant reading; an understanding of underpinning themes; the ability to collect data and evidence systematically; the ability to interpret, analyse and evaluate data and evidence; the ability to present data and evidence accurately and appropriately; critical thinking—raise and discuss issues, not just present findings; and the ability to report effectively.

The action plan for a dissertation: (1) decide on a possible focus/title and discuss with supervisor; (2) decide on research methods; (3) draw up a schedule: include completion dates for different stages; (4) organize practicalities: equipment and access; (5) set up project and collect data; (6) sort/study the data; (7) analyze/interpret the data; (8) prepare outline structure for writing up; (9) write the draft; (10) edit/check with supervisor and/or critical friend; (11) edit and finalize; (12) submit final report.

The structure of a dissertation is more or less similar to that of a research paper. However, it is elaborated and contents are furnished in divided chapters. A dissertation includes the following structural elements in its prefatory part: title page, declaration/certificate, acknowledgements, table of contents, abstract, content chapters (introduction/literature review, data and methodology, results and discussion, conclusions and recommendations), appendices, and references.

7.4 Thesis

Thesis or doctoral thesis is a long research report. The report concerns a problem or series of problems in a specific area of research and it should describe what was known about it previously, what the researcher did towards solving the stated problem(s), what the results mean, and where or how further progress in the field can be made. A thesis must make an original contribution to domain of knowledge; the research documented in a thesis must discover something hitherto unknown. Remember, the author should write to make the topic clear to a reader who has not spent years thinking about it (the author did it). The thesis will also be used as a scientific report and consulted by future workers in the laboratory who will want to know, in detail, what the author did in his/her research.

The basic structure of thesis is similar to a dissertation; however, it is more comprehensive and more informative. Results and discussion may be combined in several chapters of a thesis. The logic of the presentation is important rather than a formal structure. However, the basic structure of a thesis is being discussed in the following text.

Title Page This may vary among institutions; for example: Title, author, *A thesis submitted for the degree of Doctor of Philosophy in the Faculty of Science, under the guidance of Prof. ABC and Prof. XYZ, The University of PQR*, date. Refer Sect. 7.2.1 for writing appropriate title.

Declaration/Certificate The wording may vary among institutions, as an example of declaration: "I declare that the work contained in this thesis is original and has been done by me under the guidance of my supervisors; the work (partly or fully) has not been submitted to any other Institute for any degree or diploma; whenever I have used materials (data, theoretical analysis, figures, and text) from other sources, I have given due credit to them by citing them in the text of the thesis and giving their details in the list of references (signature/name/date)". Certificate appears something like: "This is to certify that the thesis on the topic… submitted by… embodies his/her original work supervised by me. (signature/name/date)".

Acknowledgements Most thesis authors put in a page of thanks to those who have helped them in scientific matters, and also indirectly by providing such essentials as food, educational resources, genes, money, help, advice, inspiration, friendship etc. Do not forget to acknowledge the authors of literature that have been referred during the research.

Table of Contents It helps to have the subheadings of each chapter, as well as the chapter titles. Remember that the thesis may be used as a reference in the lab, so it helps to be able to find things easily.

Abstract Refer Sect. 7.2.3 for writing a good abstract. Abstract should be a distillation of the thesis: a concise description of the problem(s) addressed, the method(s) of solving it/them, the results and conclusions. An abstract must be self-contained. Usually it does not contain references. When a reference is necessary, the relevant details should be included in the text of the abstract. Checking the word limit is important.

Introduction What is the topic and why is it important? Statement of the problem(s) is required. Everything, in this chapter, should be written in very general sense. This chapter should act as foundation of the reader to get into the complex research. Introduction should be interesting; otherwise the reader may not read the whole thesis. This chapter generally addresses the following topics, research background, significance, objectives, research questions, and methodological overview. Some of the writers prefer to provide a glimpse of contents of the chapters at the end.

Literature Review Where did the problem come from? What is already known to the problem? What other methods have been tried to solve it? Answering these questions from the existing literature can make this chapter well. How many papers should we refer; or how relevant do they have? Well, this is a matter of judgment. About a hundred is reasonable, but it will depend on the field and topic. The researcher is the world expert on the chosen topic; he/she must demonstrate it.

Materials and Methods In some thesis, it is necessary to establish some theory to describe the experimental techniques, then to report what was done on several different problems or different stages of the problem, and then finally to present a model or a new theory based on the new work. For such a thesis, the chapter headings might be: *Theory*; *Materials and Methods* {first problem}, {second problem}, {third problem}, {proposed theory/model}; and then *Conclusions*. For other theses, it might be appropriate to discuss different techniques in different chapters rather than to have a single Materials and Methods chapter.

This chapter varies enormously from thesis to thesis and may be absent in theoretical theses. However, in general sense, it is the 'description' of the research that has been carried out by the researcher. It should be possible for another competent researcher to reproduce exactly the same output by following this description. Therefore, it should be written for the benefit of the future researchers. In some theses, particularly multi-disciplinary or developmental ones, there may be more than one such chapter. In this case, the different disciplines should be indicated in the chapter titles.

Theory If the thesis belongs to basic science category, the author should include one chapter for discussing the basic theory on which the thesis is built. Of course this chapter is not exclusive for basic science subjects alone. For example when you write a thesis on any applied remote sensing topic, the author can as well include this section explaining the basic concepts and theories involved in the research.

Results and Discussion The results and discussion are very often combined in theses. This is sensible because of the length of a thesis: the thesis may have several chapters of results, and, if the author waits till they are all presented before beginning the discussion, the reader may have difficulty remembering what are talking about. The division of Results and Discussion material into chapters is usually best done according to subject matter.

In most cases, results need discussion. What do they mean? How do they fit into the existing body of knowledge? Are they consistent with current theories? Do they give new insights? Do they suggest new theories or mechanisms? Do they have any limitation? Does it have any implications that do not relate to the questions that the researcher set out to answer?

Conclusions and Recommendations Section 7.2.9 explains how to draw conclusions. It is often the case with scientific investigations that more questions than answers are produced. Does the work suggest any interesting further avenues?

Are there ways in which the work could be improved by future researchers? What are the practical implications of the work? Usually, this chapter should be reasonably short—perhaps a few pages. It is recommended to write a separate section at the end of the chapter containing author's contribution. This should be written clearly, concisely, and as a series of numbered points. This section is important to impress the examiner to obtain the degree.

References It is important to include all the important sources that have been consulted, used, or quoted in the thesis. Use any of the standard formats discussed under Research Papers. Refer Sect. 7.5 for discussion on referencing style.

Appendices If there is material that should be in the thesis but which would break up the flow or bore the reader unbearably, it is better to include it as an appendix. Some things which are typically included in appendices are: important and original computer programs, data files that are too large to be represented simply in the results chapters, forms that have been used for the collection of data, detailed census data, pictures or diagrams of results which are not important enough to keep in the main text, and so on.

Sometimes, appendices are included to document the thesis-related journal articles of which the student was the author. This is not required in every case. A list of publication related to the thesis can serve the purpose in many instances. Because, a thesis is allowed and expected to have more details than a journal article. In many cases, all of the interesting and relevant outputs can go in the thesis, and not just those which appeared in the journal. Another important issue is that the journal articles may have some common material in the introduction and material/methods sections.

7.5 Referencing Style

All the sources that have been used should be listed alphabetically at the end (list of references), and cited within the text. The author should show the readers where they found each piece of information that they have used by citing in the text. These textual citations allow the reader to refer to the list of references for further information. Various styles are available for documenting the sources within the text and in the list of works cited. However, the documentation styles laid out by the *Modern Language Association* (MLA) and *American Psychological Association* (APA) are the most common ones adopted by books, journals, and magazines. Students in Humanities courses are usually asked to follow the MLA guidelines. Students in science and technology fields are usually asked to follow the APA guidelines. In the field of remote sensing, APA is most widely used. Therefore, the APA style will only be discussed.

In APA style, the sources in a paper are listed alphabetically on a separate page headed 'References'. It follows the final page of the text and is numbered. Entries appear in alphabetical order according to the last name of the author; two or more works by the same author appear in chronological order by date of publication.

When there are two or more books or articles by the same author, repeat the name of the author in each entry. When using the examples below, it is important to follow the suggested pattern closely, even to the spacing of periods, commas, etc. It is necessary to remember that two journals following APA format may have slight differences in referencing style. Therefore, it is essential to read the style of a journal before writing the references. However, a reference written in standard APA format is generally converted into their own style by the journal typesetters. Often in many books and journals, for example this book, a simplified style is used for quick and easy referencing. These simplified styles can also be followed if accepted by the editors.

7.5.1 Single-Author Book

Bhatta, B. (2008). *Remote Sensing and GIS*. New Delhi: Oxford University Press.
<last name of the author>, <initials of first and middle name> (<year>). <*book title*>. <place of publication>: <publisher>.

If the year of publication is not indicated in the front material of the book, use the most recent copyright date. Sometimes, total page count of the book is also provided at the last as follows:

Bhatta, B. (2008). *Remote Sensing and GIS*. New Delhi: Oxford University Press, pp. 720.

Instead of pp. 720, some journals write it as p. 720. This actually denotes total page count.

This can be cited in the text as (Bhatta 2008) or (Bhatta, 2008) or "Bhatta (2008) stated …".

Leave off any titles or degrees associated with a name (PhD, Sir, or even Saint). A 'Jr.' or 'III' goes after the full name and is enclosed in commas: Arnold, C.L., Jr., & Gibbons, J.C. (1996). Impervious ….

7.5.2 Book with More Than One Author

Natarajan, R., & Chaturvedi, R. (1983). *Geology of the Indian Ocean*. Hartford, CT: University of Hartford Press.
Hesen, J., Carpenter, K., Moriber, H., & Milsop, A. (1983). *Computers in the Business World*. Hartford, CT: Capital Press.

The abbreviation 'et al.' (for 'and others') is not used in the reference list, regardless of the number of authors, although it can be used in the parenthetical citation of material with three to five authors (after the initial citation, when all are listed) and in all parenthetical citations of material with six or more authors. In case of two authors it is always cited as (Natarajan & Chaturvedi, 1983). In case of three to five authors, at the first citation it is written as (Hesen, Carpenter, Moriber,

& Milsop, 1983), and thereafter as (Hesen et al., 1983). For more than five authors it is always cited as (<Author> et al., <year>). However, because of this complexity, many journals always use 'et al.' for more than two authors.

7.5.3 Edition Other Than First

Bhatta, B. (2011). *Remote Sensing and GIS* (2nd ed.). New Delhi: Oxford University Press.

7.5.4 Edited Volume

Stanton, D. C. (Ed.). (1987). *The Female Autograph: Theory and Practice of Autobiography*. Chicago: University of Chicago Press.

7.5.5 Book Without Author or Editor Listed

If the reference is to a book without any authorship, the title of the book takes the place of the author.

Webster's New Collegiate Dictionary (1961). Springfield, MA: G. & C. Merriam.

7.5.6 Multi-Author Article/Chapter in a Multi-Author Book

Batty, M. & Howes, D. (2001). Predicting temporal patterns in urban development from remote imagery. In J.P. Donnay, M.J. Barnsley & P.A. Longley (Eds.), *Remote Sensing and Urban Analysis* (pp. 185–204). London and New York: Taylor and Francis.
<chapter author(s)> (<year>). <chapter title>. In <name of the editors; initials of first/middle names then the last name> (Eds.), <*book name*> (pp. <starting and ending page numbers of the chapter> . <place> : <publisher> .

Editors' names are not used in the citation; for the above example, citation is: (Batty & Howes, 2001).

7.5.7 Journals/Periodicals

Hasse, J. (2004). A geospatial approach to measuring new development tracts for characteristics of sprawl. *Landscape Journal*, 23(1), 52–67.
<author(s)> (<year>). <title of the article> . <*journal name*> , <volume> (<issue>), <starting and ending page numbers of the article>

Inclusive page numbers are used; the abbreviations 'p.' or 'pp' are not used.

7.5.8 Newspaper Articles

If the article is 'signed' (that is, the author's name is printed), it begins with the author's name. Notice the discontinuous pages.

> Poirot, C. (1998, March 17). HIV prevention pill goes beyond 'morning after'. *The Hartford Courant*, pp. F1, F6.

If the author's name is not available, the reference begins with the headline or title in place of the author's name.

New exam for doctor of future. (1989, March 15). *The New York Times*, B-10.

7.5.9 Online Article

> Klein, D.F. (1997). Control group in pharmacoptherapy and psychotherapy evalu-
> ations. *Treatment*, I. Retrieved November 16, 1997 from the World Wide Web:
> http://www.apa.org/treatment/voI1/97_a1.html

7.5.10 Online White Paper

> Barnes, K. B., Morgan, J. M., III, Roberge, M. C., & Lowe, S. (2001). *Sprawl development:
> Its patterns, consequences, and measurement*. A White Paper, Towson University. URL.
> http://chesapeake.towson.edu/landscape/urbansprawl/download/Sprawl_white_paper.pdf.

7.5.11 Online Conference/Seminar Proceedings

> Angel, S., Parent, J., & Civco, D. (2007). Urban sprawl metrics: an analysis of global
> urban expansion using GIS. Proceedings of ASPRS 2007 Annual Conference, Tampa,
> Florida May 7–11. URL: http://clear.uconn.edu/publications/research/tech_papers/
> Angel_et_al_ASPRS2007.pdf.

7.5.12 Multiple publications of Same Author from Same Year

> Bhatta, B. (2009a). Analysis of urban growth pattern using remote sensing and GIS:
> a case study of Kolkata, India. *International Journal of Remote Sensing*, 30(18),
> 4733–4746.
> Bhatta, B. (2009b). Modeling of urban growth boundary using geoinformatics.
> *International Journal of Digital Earth*, 2(4), 359–381.

They should be cited in the text as (Bhatta, 2009a, 2009b) for both or for single (Bhatta, 2009a).

For further detail one may visit the official website of APA at http://www.apa style.org/. Although not officially linked to the authors of MLA or APA style, the following websites are from reputable colleges and offer discussions of the various styles that can supplement the discussion in this section:

http://owl.english.purdue.edu/owl/resource/560/01/
http://www.monroecc.edu/depts/library/apa.htm

7.6 Some Guidelines on Writing

It is true that writing is an innate talent, something we are born with or born without. However, it is not essential to have this talent in born—it can be practiced and learned. Thought of in this way, each of us must begin as an apprentice, learning tools and techniques, training, and hopefully eventually even perfecting our abilities. This section is intended to offer an initial training.

When we are about to begin, writing a thesis or even a research paper seems a difficult task. That is because it is a 'difficult task'. Fortunately, it will seem less daunting once we have a couple of chapters done. Towards the end, we shall even find ourselves enjoying it—an enjoyment based on satisfaction in the achievement, pleasure in the improvement in our technical writing, and of course the delight of approaching the end. Like many tasks, writing usually seems worst before we begin; so let us look at how we should make a start.

It is always recommended to make up an outline first: several pages containing chapter headings, sub-headings, some figure titles (to indicate which results go where) and perhaps some other notes and comments. Once a list of chapters (and under each chapter heading a reasonably complete list of things to be reported or explained) is prepared, we start writing. Remember, this writing is an initial draft that requires going through several revisions. This initial draft is called *sloppy first draft* or even *zero draft* (meaning the one before the first draft; see Becker 1986 and Bolker 1998).

When we sit to type, our aim should not be a thesis or a research paper—but something simpler. Our aim is just to write a paragraph or section about one of our sub-headings. For some of us, beginning on page one remains difficult—especially at first, when we often do not yet exactly know what we are going to say. Therefore, it is always preferred to start with an easy one; this gets us into the habit of writing and gives us the confidence to proceed further. Never mind if it turns out to be in the middle of page eight when the paper is finished; word-processing software has eliminated any concerns about that. Often the *Materials and Methods* part is the easiest to write; because, this is the new thing that has been done by the researcher—she/he is the world expert in this. Somewhere in the middle of literature review can also be started. Writing should be careful, formal, and in a logical order.

Most students new to research fear they have little flexibility in the writing of their papers/thesis; it does not have to be that way. One of the most important tools in taking charge of the story the researcher are about to tell is writing a working outline. How do we make an outline of a chapter? Such an outline may be a series of notes and reminders. It is not written for someone else; therefore need not to be formal. In fact, outline for this chapter consisted only of twenty five short words or phrases designed to remind of the things to write about, and to suggest an order to those topics. Materials in the outline are of free choice. Beginners may create a more detailed outline. Assemble all the results/figures that will be used in it and put them in the order that would be used. One might as well rehearse explaining it to someone else—after all every researcher probably gives several seminars based on the research. Once the most logical order has been finalized, note down the key words of the explanation. These key words provide a skeleton for much of the chapter outline. Once the outline is ready, discuss it with the supervisor or guide. This step is important: she/he will have useful suggestions, but it also serves notice that she/he can expect a steady flow of chapter drafts that will make high priority demands on her/his time. Once the supervisors have agreed on a logical structure, the writing can be started.

It is encouraging and helpful to start a filing system. Open a word-processor file for each chapter and one for the references. We can put notes as well as text in these files. While doing something for Chapter n, we may think "Oh I must refer back to/discuss this in Chapter m" and so we must put a note to do so in the file for Chapter m. Or we may think of something interesting or relevant for that chapter. Keep a back-up of these files and do so every day at least. One may also use a physical filing system: a collection of folders with chapter numbers on them. This will make you feel good about getting started and also help clean up the desk. These files will contain not just the plots of results and pages of calculations, but all sorts of old notes, references, calibration curves, suppliers' addresses, specifications, speculations, letters from colleagues, or anything relevant to that chapter.

One should consult the supervisor and make up a timetable for writing a thesis: a list of dates on which the first and second drafts of each chapter will be prepared. This structures the time and provides intermediate targets. One may want to make the timetable into a chart with items. It becomes very helpful if we can mandatorily write two or three pages every day. If one day is skipped, try to write double in the next day. This approach, although hard to maintain, is very useful for the completion within time, whether it is an article or thesis or a book. When all else fails, sometimes writing under pressure is just what we need: with the minutes ticking away until the date is due we may find that we have to write something. While this is not the best strategy for creating truly well, however, nearly every writer has experienced its effectiveness. Even the most seasoned scholar who begins her/his work long before a deadline is near, may find her/him-self with much left to be written as the deadline looms. It is obvious that the knowledge generated from research should be quickly publicized in referred journals to avoid wasteful duplication of work and to establish the researcher's claim to the priority of discovery. This must be done well before the thesis is being written. These journal articles indeed help a lot in framing and writing a thesis.

Whenever we sit to write, it is very important to write something. So write something, even if it is just a set of notes or a few paragraphs of text that would not be shown to anyone else. It would be nice if clear, precise prose leapt easily from the keyboard, but it usually does not. Most of us find it easier to improve something that is already written than to produce new. So put down a draft (as rough as one like) for own purposes, then clean it up. Supervisors generally read each chapter in draft form. She/he then returns it with suggestions and comments. One should not be upset if a chapter, especially the first one, returns covered in red ink. Supervisors want the thesis to be as good as possible, because of his/her reputation as well as the scholar. Scientific writing is a difficult art, and it takes a while to learn. As a consequence, there will be many ways in which the first draft can be improved. So take a positive attitude; each comment of the supervisor tells a way in which the thesis can be made better. Even for native speakers of English who write very well in other styles, one notices an enormous improvement in the final chapters comparing the initial drafts.

In order to engage our creativity, and to recognize it, for us as we write and for our readers as they read, academic writers use many of the same techniques that journalists and fiction writers do. Some use metaphors and descriptive language to help evoke the feeling of a scene, a place, or a person. What is most important here is that we should write in a style that is comfortable to us. Many writers of research papers attempt to imitate the often stodgy style of much academic writing, choosing words that sound 'harder' or more complex (and are often longer) rather than using the 'regular' words of everyday language. New researchers often mistakenly take, not the best, but the worst prose (Williams 1995). The text must be clear. Good grammar, simple words, smaller sentences, and thoughtful writing will make the paper/thesis easier to read. Scientific writing has to be a little formal—more formal than a newspaper article or a textbook. Native English speakers should remember that scientific English is an international language. Slang and informal writing will be harder for a non-native speaker to understand.

Short, simple phrases and words are often better than long ones. On the other hand, there will be times when complicated sentence is needed because the idea is complicated. Some lengthy technical words will also be necessary in many cases. Sometimes it is easier to present information and arguments as a series of numbered points, rather than as one or more long and awkward paragraphs. A list of points is usually easier to write. One should be careful not to use this presentation too much: the thesis must be a connected, convincing argument, not just a list of facts and observations. One problem for academic prose is that, in our commitment to 'telling the facts', we may get lost in a less-than-creative (or even boring!) way of presenting those 'facts' (DeLyser 2010). Yet, no matter how clear our evidence may be, each research paper does much more than simply presenting the 'facts' about a particular topic. Although research papers do tell facts, those facts are always shaded (to different degrees) by the circumstances and the person doing the research and writing (Becker 1986). Each research paper then, by its very nature, grants its author the opportunity to 'tell the story' in a myriad of ways.

One important stylistic choice is between the active voice and the passive voice. The active voice ("I classified the image") is simpler rather than passive voice ("The image was classified"), and it makes clear what we did and what was done by others. Most of the teachers and authors of technical communication shout for active voice. "Sorting the wheat from the chaff (in case of passive voice) ... can be difficult, so a cautious writer will stick to a simple and direct style, one that will make your ideas clear to readers and runs not the risk of imitating an embarrassing example of scholarly turgidity. Read your prose carefully back to yourself, and listen for your own voice in it. Strive to recognize yourself, not someone else, in your own writing" (DeLyser 2010). The passive voice makes it easier to write ungrammatical or awkward sentences (Raman and Sharma 2004). "If you use the passive voice, be especially wary of dangling participles" (Raman and Sharma 2004). However, people generally avoid active voice in a thesis or scientific writing because of two reasons: (1) most of the scientific writings are written in the passive voice, and (2) some journal editors and teachers do not accept papers/thesis in active voice. Further, if one chooses the active voice, then also first person singular and second person singular/plural are generally avoided, because, the use of 'I' and 'you' is considered to be immodest. Consider the following sentence: "as one can see, this book has been written mostly in passive voice; in case of active voice, first person singular and second person singular/plural have been avoided (as you can see, I have written this book mostly in passive voice; in case of active voice, I have avoided first person singular and second person singular/plural)". Although the later is easier, clearer, and more effective to communicate, is it decent enough in respect of a formal scientific writing? However, there is no harm if one uses active voice, first person singular, and second person singular/plural when reporting a work that has been done by her/himself.

Another important thing is the presentation. In general, students spend too much time on diagrams, font styles, decoration, etc.—time that could have been spent on examining the arguments, making the explanations clearer, thinking more about the significance, and checking for errors in the algebra. The reason, of course, is that decoration is easier than thinking. Decoration is absolutely not required in a scientific literature; rather, it should be decent and simple—as simple as a journal article. Over-decorative writing may not be taken seriously by the examiner/reader. Next comes the length. There is no strong correlation between length and quality. There is no need to leave big gaps to make the thesis thicker. Further, readers will not appreciate large amounts of vague or unnecessary text. High volume, in general, discourages the reader.

Finally, the most important thing is the arguments. When not thinking about writing, the word 'argument' conjures the image of a fight or a dispute. In scholarly writing however, arguments are the key to a successful research; each research paper/thesis makes an argument: it states a case, and presents evidence to support that case. Though some academic arguments (polemics) are forceful, and may actually feel like a fight (they may, for example, take on the work of another scholar, refuting it as incorrect), others can be far more subtle, or complex. A good research paper may even present an argument showing multiple sides of an issue—the argument may be that each side is valid, that each side has its merits.

We must gather and assemble our evidence in support of whichever argument we have chosen, whether using statistics and tables, references to published works, quotes from interviews, photographs, or all of the above. Of course, we cannot simply exclude evidence that may seem to contradict our arguments. In fact, by including and addressing such claims, we may actually be able to make our arguments stronger. In making sincere efforts to explain and engage all we need to, we may find that our argument has at times branched away from its focus. If that is the case, it is recommended to consider moving the branching portions of the argument to one or more endnotes or footnotes.

Once the writing is complete, it is better to leave it for a while—few days, if possible for few weeks and even months. Just forget everything. Then start the revision. Real revisions involve much more than running a spellchecker—the real point of revision is not just correction but improvement. Polishing of language, style, arguments, and structure are very important. Remember, still there is some room left to do something, e.g., new argument, new figure, new table, and so on. It is essential to proofread every word on every page closely and carefully, literally listening to each word in turn. It often helps in catching small mistakes to actually read the paper aloud. It is often helpful to have someone (other than the author(s) themselves) to read some sections of the paper/thesis, particularly the introduction and conclusion parts. It may also be appropriate to ask friends to read some sections of the paper/thesis which they may find relevant or of interest, as they may be able to make valuable contributions. In either case, only give them revised versions, so that they do not waste time correcting the grammar, spelling, poor construction, or presentation.

In looking back over this section, one may notice that, writing is 'very' difficult and tedious. Indeed, writing is a difficult task, especially to the new researchers; but it is not 'so' difficult. Further, it is not tedious at all. The writing process itself is formative: we actually come up with, construct, and develop our ideas while we are writing. Our ideas occur to us and change literally as we write, and those ideas (along with the ultimate finished paper/thesis) are a product not just of the data that is 'out there', but also of how we found that data, how we approached the topic, and even who we are (writing gives the appearance and importance of the author). Thus, "the process of writing a research paper needs to be seen as integral to the research itself" (DeLyser 2010).

One may refer following literature for further reading:

Ballenger (2008): Shows how good research and interesting writing work together in compelling research papers, offered in a step-by-step approach to focus a topic and get the paper finished.

Cook (1985): Rich examples teach editing techniques and strategies designed to help writers see the weaknesses they may otherwise miss in their own prose.

DeLyser and Pawson (2010): Offers strategies and techniques for communicating research to different audiences in different forms—in writing, in verbal presentations, on websites, etc.

Lester (2009): A comprehensive guide to traditional- and electronic-based research and presentation, this text also navigates the subtleties of different citation styles and methods of documentation, and answers questions about grammar and usage.

Turabian (2007): A classic text offering both a style guide and a research guide, now updated to include extensive material covering on-line sources.

Veit (2004): Uses examples from actual research papers to help students maintain a focus on the goals of writing research papers, explaining the research and writing processes in a step-by-step manner, and showing along the way how conventional writing rules and formats help writers achieve those goals efficiently.

References

Ballenger BP (2008) The curious researcher: a guide to writing research papers, 6th edn. Longman, New York

Becker HS (1986) Writing for social scientists: how to start and finish your thesis, book, or article. University of Chicago Press, Chicago

Bolker J (1998) Writing your dissertation in fifteen minutes a day: a guide to starting, revising, and finishing your doctoral thesis. H. Holt, New York

Cook CK (1985) Line by line: how to edit your own writing. Houghton Mifflin Co., New York

DeLyser D (2010) Writing it up. In: Gomez B, Jones JP III (eds) Research methods in geography: a critical introduction. Blackwell Publishing, West Sussex, pp 424–436

DeLyser D, Pawson E (2010) From personal to public: communicating qualitative research for public consumption. In: Hay I (ed) Qualitative research methods in human geography, 3rd edn. Oxford University Press, Melbourne, pp 266–274

Lester JD (2009) Writing research papers: a complete guide, 13th edn. Longman Publishers, New York

Raman M, Sharma S (2004) Technical communications: principles and practice. Oxford University Press, New Delhi

Turabian KL (2007) Student's guide for writing college papers, 7th edn. University of Chicago Press, Chicago

Veit R (2004) Research: the student's guide to writing research papers, 4th edn. Longman Publishers, New York

Williams J (1995) Style: toward clarity and grace. University of Chicago Press, Chicago

Index

B. Bhatta, *Research Methods in Remote Sensing*, SpringerBriefs in Earth Sciences,
DOI: 10.1007/978-94-007-6594-8, © The Author(s) 2013